Emerging Topics and Questions in Infocommunication Technologies

Emerging Topics and Questions in Infocommunication Technologies

Edited by

Sergey Balandin, Vladimir Deart and Valtteri Niemi

Cambridge Scholars Publishing

Emerging Topics and Questions in Infocommunication Technologies

Edited by Sergey Balandin, Vladimir Deart and Valtteri Niemi

This book first published 2020

Cambridge Scholars Publishing

Lady Stephenson Library, Newcastle upon Tyne, NE6 2PA, UK

British Library Cataloguing in Publication Data
A catalogue record for this book is available from the British Library

ISBN (10): 1-5275-5413-9
ISBN (13): 978-1-5275-5413-9

TABLE OF CONTENTS

PREFACE

This collection consists of selected papers from the IEEE FRUCT 2019 Conferences. The conference is a regular and significant open event organized by the Open Innovations Association FRUCT, which has a long and positive history of pioneering and assisting long-term cross-border pan-European academic and commercial cooperation in the field of Information and Communication Technology (ICT). In total, the conference received over 300 submissions; of these, less than 130 were accepted for publication as full papers (with an acceptance rate less than 40%). For this collection of essays, we selected 14 papers. These are extended versions of the best papers from the conference.

The main motivation for organizing regular FRUCT conferences is the distinctive features of modern IT and ICT industries, which are experiencing a dramatic shortening of the innovation cycle and a reduction of the period before technology becomes commercially viable. This is due to competition between market players pushing all manufacturers to accelerate innovations combined with a speeding up of technological progress caused by the growing expansion of intellectual resources invested in R&D and design activities. This trend is a key challenge for leading educational and research institutions around the globe. In FRUCT we believe that it is strategically important to develop pan-European cooperation and cultural exchange (and also keep Russian scientists in the loop). FRUCT aims to help regional teams in following up and adopting university research and education and aligning it to real industrial challenges. The primary mission of the conference is to strengthen cooperation between the Russian and Finnish academic communities, increase the visibility of research teams, and build up direct personal contacts between academic and industrial experts.

FRUCT conference activities are based around the idea that continuous development of the strategic partnership between industrial and academic research is a key factor in the success of the modern innovation ecosystem. There are a few success stories of such strategic partnership frameworks functioning in different parts of the world. These programs bring significant benefits to the parties involved and fuel their R&D. As we know, the fundamental science work undertaken by universities and other academic organizations should not be directly attached to existing industries, but industrial research can benefit greatly by early access to results and information about important trends and weaker pathways. At the same time, many universities are also active in applied research, but to be efficient they need timely feedback from industry. Another key driver for ensuring stronger connections between academia and industry is the shortening of the innovation cycle. An interesting new trend for addressing this need is the building of open innovation frameworks targeting the development of strategic partnerships between industrial and academic research. Such framework programs help potential research partners find each other and jointly incubate new competencies.

The FRUCT Association was established at the beginning of 2007, when the need for such a community was recognized by a group of enthusiasts and supported by Nokia's university cooperation program in Russia, Saint-Petersburg State University of Aerospace Instrumentation, and the University of Turku. Over the past few years, more than 30 research groups from top universities in Finland, Russia, Italy, Ukraine, Denmark and India, as well as the R&D institution of the Russian Academy of Science, have joined the FRUCT Association. Nowadays, FRUCT is one of the largest and most significant associations facilitating the development of research and collaboration between universities and ICT companies in the region. The FRUCT Association has good contacts with industry and events are attended by many representatives of industrial companies. FRUCT has a good history of cooperation with such giants as Nokia, Intel, Samsung, Dell EMC, MariaDB, and Microsoft. Formal membership of the FRUCT Association is by invitation, but FRUCT is an open society and we welcome all interested universities and companies to engage in cooperation and join our friendly community.

The main objective of FRUCT is to establish and develop contacts between academic and industrial research teams from different countries, especially in Russia, Finland, India, Italy, Ukraine, and other European countries. The association is developing processes to enable the early involvement of undergraduates and postgraduates in scientific activities, creating joint teams for solving challenging scientific problems with the use of knowledge intensive technologies and increasing the prestige of scientific and research work. FRUCT has developed many processes that support mutually beneficial cooperation and the development of strategic partnerships between academic and industrial research. The processes developed help overcome barriers in the path of innovation and illustrate how business can benefit from greater social responsibility and the support of long-term research and academic cooperation.

The essays in this collection were selected on the basis of reviewer comments and conference participant feedback. The contents has been organized into four chapters that best highlight the most popular conference sections addressing the hottest topics in ICT research and development.

The first chapter is titled Next Generation Networks. Today, most telecommunications companies use NGN technology for providing their subscribers with a Triple Service (Data, Voice, and Video). Providers have received good feedback from users in terms of the Quality of Experience (QoE) of data transmission, voice perception, and video streaming, all depending on the Quality of Service (QoS) parameters of the existing network. As video streaming now uses High Definition and Ultra-High Definition it needs 10 times the bandwidth of Standard Definition. Providing high resolution video with the required QoE for users with expanding bandwidth demand is becoming an urgent task. The global trend of modernization of NGNs involves software defined networks (SDNs), which can resolve the problem of ensuring the required quality for different types of media streaming. But SDNs must ensure an adequate level of reliability, which is no simple task for centrally controlled networks. The ever-increasing number of new applications for NGNs is a real trend and serves as an attractive point for new customers.

The second chapter discusses Innovative Knowledge-based Systems. This chapter focuses on the development problem of innovative knowledge-based systems. Exciting examples of such system classes were presented and discussed at the IEEE FRUCT 2019 Conferences. The problem domain topics include: augmented reality systems for industrial needs; management systems for the Internet of things; information spaces for smart factories; monitoring and analytics for agricultural data; mobile video and interactive applications for everyday use; ontology-oriented software development for smart spaces; and time-series analysis for recommendation systems. The discussed topics meet new research challenges and the authors define relevant problems, suggest possible formalizations and concept models, and illustrate the study with demonstration implementations.

The third chapter present research on Innovative Applications and Services. Innovative applications play a significant role in modern life.

Robotics, value added services, and intelligent mobile applications are now widespread in our daily lives. Nowadays, the functionality of the modern smartphone goes far beyond making calls. Smartphone devices allow the implementation of calculations and proactively provide the user with multi-modal information, as well as connecting him/her to the e-world in any physical location. Modern information and communication technologies provide a strong basis for research and are widely used by the research community. This chapter presents papers related to research methodology development, as well as prototypes for the implementation innovative application development.

The last chapter focuses on Innovations in Healthcare and eHealth. Visual analysis and interpretation still represent the gold standard in clinical practice. Therefore, the success of correct clinical evaluation, depends on an examiner's skills and is prone to subjective variability. In the age of personalized medicine, more objective methods are needed to improve diagnosis, characterize diseases, make prognoses, and monitor therapy. Image analysis and processing constitute essential tools for providing non-invasive or minimally invasive alternatives to improve, quantify, and maximize *in vivo* information obtainable in clinical practice, thus playing a fundamental role in clinical decision-making.

April 2020

Sergey Balandin
Vladimir Deart
Valtteri Niemi

CHAPTER 1.1

TCP ALTERNATIVES FOR VIDEO STREAMING VIA OTT

ALEXEY NESTERKIN, VLADIMIR DEART

Abstract. Video services using over-the-top (OTT) delivery are becoming extremely popular across the world, but video streaming through an unguaranteed Internet path can cause a lot of problems in terms of maintaining a suitable quality of experience (QoE) level. Previous work has addressed the influence that quality of service (QoS) parameters have on the re-buffering effect. Here we continue this research for video at higher resolutions of 1080p, 2K, or even 4K. The primary purpose of this research is to investigate how delay and packet loss influence the re-buffering effect in the HLS, RTMP, and MPEG-DASH streaming protocols during the delivery of high-resolution video.

Introduction

Over-the-Top (OTT) technology providing video-on-demand services is becoming more and more popular in countries all over the world. The number of users of streaming video services is increasing significantly from year to year. The biggest providers have already implemented this technology in their IP networks to achieve larger profits from greater utilization of the existing network infrastructure. This has all lead to an increase in competition between providers seeking to capture the market.

Forecast analysis of OTT market trends suggest a continuing increase in the volume of OTT video traffic and the use of smartphones. An analysis of the Internet up to 2022 by Cisco Systems [1] suggests that we will see a decrease in the use of personal computers to access the Internet (PC): PC activity is predicted to decrease from 49 % to 19 % (Fig. 1). At the same time, smartphones are becoming more and more ubiquitous Internet devices and will account for 50 % of total traffic on the Internet by 2022 (Fig. 1).

CHAPTER 1

NEXT GENERATION NETWORKS

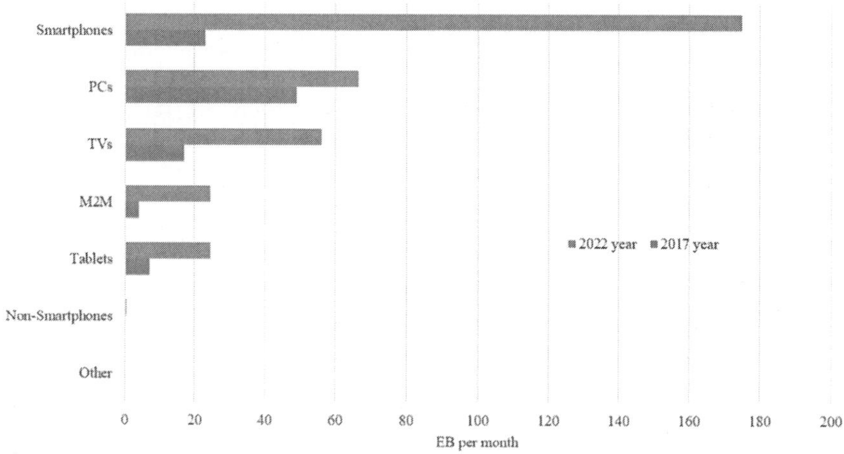

Fig. 1. Most active Internet devices

Currently, a primary use of smartphones and tablets is viewing video [2]. As a result, the volume of IP video traffic will also increase year by year. By 2022, IP video traffic is set to account for 82 % of traffic [1] (Fig. 2).

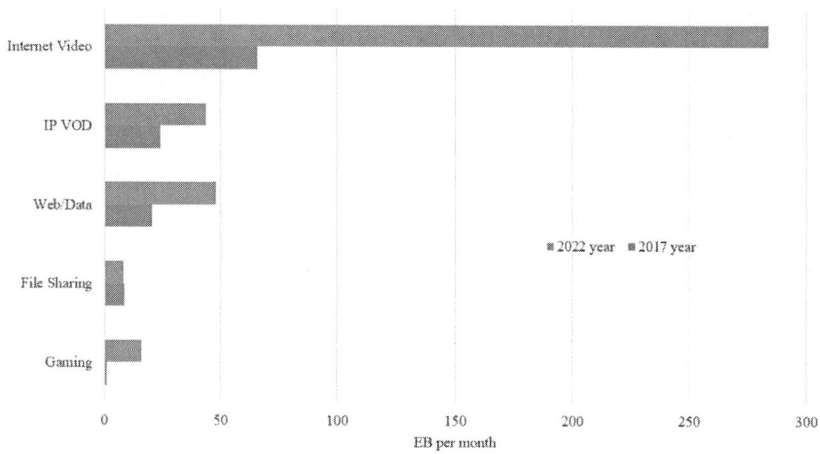

Fig. 2. Most active Internet application

Competition for customers is forcing OTT streaming video providers to offer new and varied services. Most of them offer video with 1280 × 720 HD (High-Definition) or 1920 × 1080 FHD (Full High Definition) resolution, but this is no longer enough. Consumers are seeking even better quality video at 3840 × 2160 UHD (Ultra-High Definition) 4K, or even 7680 × 4320 UHD 8K resolution. This trend has been identified in TV technology research [3].

This will increase Internet traffic because videos in UHD resolution require more than twice the bandwidth of videos in FHD, and more than ten times the bandwidth of those in SD (Standard Definition).

The analysis made by Cisco Systems [1] suggests that consumers will give preference to HD and 4K UHD in IP VoD (video on demand) services (Fig. 3).

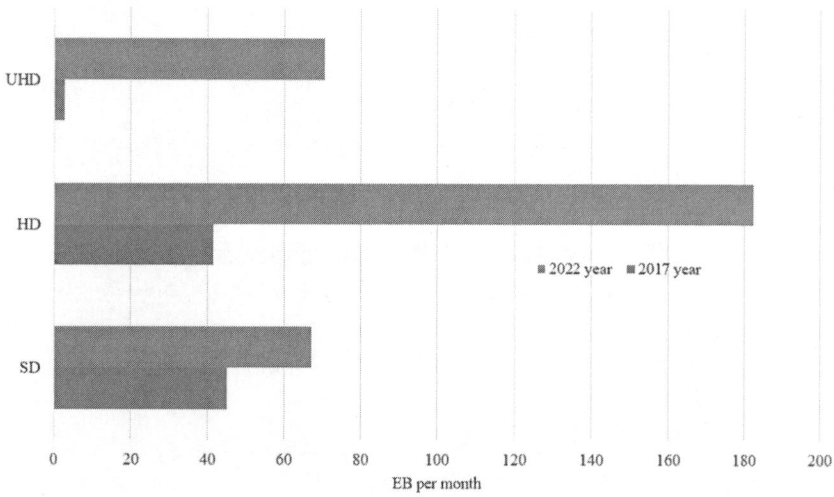

Fig. 3. Global IP video traffic

These trends in Internet traffic will change its overall pattern. The role of content delivery networks (CDNs) is set to increase significantly. According to the analysis in [1], these services will carry 72 % of total Internet traffic by 2022 (Fig. 4). As such, OTT services will be located closer to the end user, the consumer, decreasing delivery latency during OTT video streaming.

Some modern devices, such as smartphones, tablets, and TVs, support 7680 × 4320 8K UHD resolution. The forthcoming 5G network will allow all consumer devices to have a high-speed Internet connection. This new challenge creates demand for new services from OTT streaming service providers [3] who must be able to supply users with video at the maximum resolution of their devices: 3840 × 2160 4K or 7680 × 4320 8K.

Our previous research into video streaming with a resolution of 1280 x 720 and a mean bitrate value of 5.68 Mbit/s showed that the combination of delay and loss lead to significant degradation of the QoE [4]. The re-buffering effect appears with packet loss of 0.5 % and delays of 12-15 ms for different streaming protocols. Most client devices use wireless technology to access the Internet over the final mile with loss rates and delay.

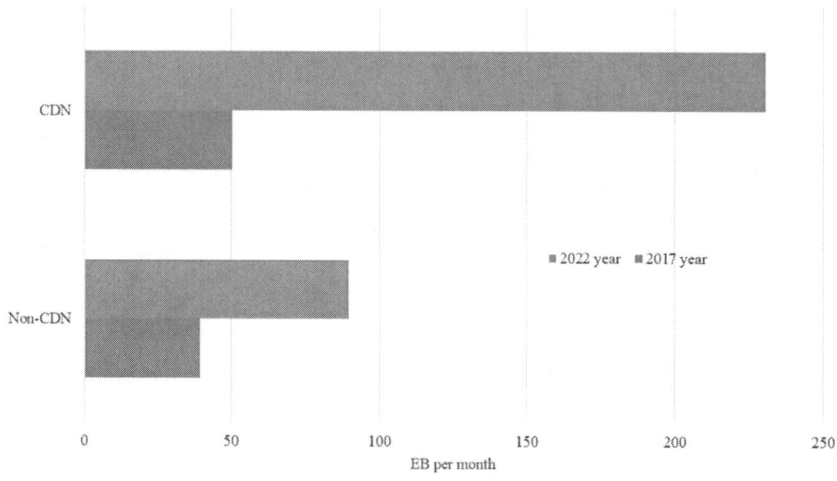

Fig. 4. The role of CDNs in content delivery

This research is devoted to the estimation of packet loss and delay thresholds in the re-buffering effect for the HLS (HTTP live streaming), RTMP (real-time messaging protocol), and MPEG-DASH (MPEG dynamic adaptive streaming over HTTP) streaming technologies for 1920 × 1080 HD video. An assessment was made of the significant influence of packet loss and delay on re-buffering. A proposal is suggested for acceptable levels of delay and packet loss for streaming video with 7680 × 4320 8K UHD resolution, excluding re-buffering.

This paper is structured as follows. In the following section, current research on the problem is reviewed. In the next section, the experimental platform is described and the characteristics of the video clips examined are presented. In the fourth section, the results are presented. In the fifth section, the influence of packet loss and packet delays on re-buffering for different video resolutions are analyzed and discussed. The sixth section describes different transport protocols, which can be used as alternatives to TCP. The seventh section offers some conclusions and the final section is devoted to future work.

Review of OTT Technology Research

A. Consumer satisfaction with OTT technologies

In the early days of OTT implementation, consumers were very tolerant of re-buffering and delayed starts to their videos. The extension of OTT video services has led to an increase in consumer expectations. Consumers now wish to watch video content anywhere and without any degradation: this new capability can be summed up as "I can watch what I want where I want".

Here are some examples from consumer satisfaction research done by various companies:

- "Viewers are increasingly less forgiving of re-buffering. The average number of times a viewer will let a video re-buffer before they stop watching has dropped from 2.7 times in 2016 to 2.2 times in 2018" [2].
- "OTT viewers now care more about QoE than the choice of content" [5].
- "A total of 81 % of consumers experience buffering (a lag that disrupts playback while content is in-progress) or delayed video start (when a video takes significant time to initially load), according to this year's findings" [6].
- "Video buffering causes a decrease in positive emotions (happiness down 14 %), a 9 % increase in disgust, a 7 % increase in sadness, and an 8 % decrease in focus. Taken together, these represent an overall 16 % increase in negative emotions" [7].
- "Today's promiscuous viewers will not tolerate any buffering. Intermittent outages and delays are out of the question. Just a few seconds is all it takes for them to switch off and move over to a competitor's service" [3].

Taking into account the user opinions, it is possible to highlight three common problems, emphasizing:

- The re-buffering effect;
- Delayed start;
- Reduced video bitrate.

We are also continuing to study consumer QoE. Year by year, customers become less tolerant of the re-buffering effect and variable video quality. They do not want to have any distortion while watching an OTT video stream. Based on this trend, we have set a new focus for our research of finding the threshold of QoS network parameters past which we will see the re-buffering effect for different video resolution and streaming technologies.

B. Results of previous research

Competition is always good for users, and, as such, OTT streaming video providers must deliver videos with the best QoE possible. For this reason, a lot of research has been done by authors to evaluate network performance factors affecting QoE. During the first phase of this research, experiments for subjective evaluation of QoE with different sets of QoS parameters (packet loss, delay, and throughput) were undertaken using an experimental platform. The results of these experiments [8, 9, 10] have shown that IPLR (IP packet loss ratio) and IPTD (IP packet transfer delay) have a significant impact on QoE. Unfortunately, these subjective measurements display great dispersion and are time-consuming to develop. For objective measurements, the authors suggest using factors like re-buffering, when the receiving buffer is empty and video playback is stopped. An estimation of video quality can be undertaken based on the frequency and mean duration of re-buffering [10].

Research was performed to discover thresholds for QoS parameters that radically influence QoE. The simultaneous influence of two QoS parameters, packet loss and delay, was shown to have a significant impact on re-buffering. In this work, three streaming technologies were compared: HDS (Adobe Flash HTTP Dynamic Streaming); RTMP; and pseudo-HTTP streaming. A video clip was chosen with a resolution of 1280 x 720 pixels; VBR (variable bit rate; the mean value of the bitrate was equal to 5.68 Mbit/s) and quick changes of scene with dynamic objects; and audio at 192 kbit/s. The experimental results for the influence of QoS parameters on re-buffering thresholds [4] are presented in Table 1.

The two top rows show the influence of these two distinct factors: packet loss and delay. The bottom rows show the simultaneous influence of these two QoS parameters.

Table 1. Re-buffering thresholds

	Streaming Technologies		
	HDS	RTMP	Pseudo-HTTP
Packet Loss, %	2.4	2.8	3.1
Delay, ms	40	34	62
Delay, ms (packet loss, 0.5 %)	12	14	15
Delay, ms (packet loss, 1 %)	9	10	11

The research shows that the thresholds after which re-buffering starts are rather high when compared to the objectives for Class 4 IP networks [11]. In real networks, both factors exist simultaneously. With a combination of the above-mentioned parameters, the thresholds for delays were 4-5 times lower when compared to the thresholds for the separate factors, being in the range of 12-15 ms for packet loss of 0.5 % and 9-11 ms for packet loss of 1 %. These values are very close to measured QoS parameters on real IP networks.

Comparison of streaming technologies was undertaken by looking at the influence of both factors in combination. Best results in terms of resistance to packet loss and delays were shown by pseudo-HTTP, while HDS was the worst. RTMP showed intermediate results, but RTMP has a specific functionality as a decoder at low bitrates, leading to a slide show, and this mode also influences a user's perception [12].

As OTT becomes more popular among national and international providers, the quality of experience for streaming technologies becomes more and more important. For IPTV (internet protocol television) technology, ITU-T Recommendation G.1080 [13] covers the most important aspects for supporting QoE. For OTT services there are currently no such regulatory documents. Meanwhile, the mechanisms for supporting QoE of streaming over OTT are becoming vital for video service providers. For practical needs, providers need a recommendation for monitoring and controlling specific QoS parameters to exclude the influence of network performance parameters on the quality of user perception.

C. Other work by the authors

Other work by other authors were oriented to the design of bitrate adaptation algorithms for adaptive bitrate streaming (ABR) schema [14, 15, 16]. Many implementations of ABR streaming exist, including HLS [17], RTMP [12], and MPEG-DASH [18]. Each has its own proprietary implementation and sees slight modifications to the basic ABR technique. Most of this research has been targeted towards an open-source standard for ABR streaming: MPEG-DASH. It is usually assumed that a lower video bitrate is present. For example, in [14] a bitrate of 800 kbps for 720 × 576 SD and 2 Mbps for 1280 × 720 HD video resolution was assumed. This assumption does not suit our work. Netflix [19] gives the following recommendations:

- 3.0 Mbps for SD video quality;
- 5.0 Mbps for HD video quality.

The focus of this research was solely on the ABR algorithm and it did not consider current market demands and trends in consumer requests. Users want to watch video with FHD 1920 × 1080 resolution and without any video quality degradation. ABR technology mitigates the re-buffering effect by using video quality degradation, but this would be unacceptable to OTT streaming users.

Experimental Platform

In this section, we describe our experimental platform and the assumptions of the network parameters and video used in the experiments. The experimental platform (Fig. 5) consists of three virtual machines:

- Ubuntu Server with installed ngnix web server [20, 21].
- Network emulator WANem [22] used to set network QoS parameters (packet loss and delay).
- Ubuntu Client with a web browser.

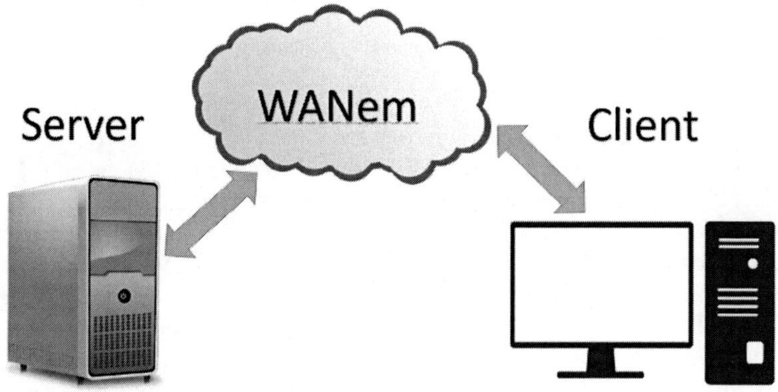

Fig. 5. Experimental platform

A full description of the experimental platform is given in Table 2.

Table 2. Detailed information about the platform

Name	Description
PC Hardware Platform	CPU - Intel Core i7 – 4770 RAM – 16 Gb Video card - GeForce 210 Gainward PCI-E
OS Server	Ubuntu 16.04.4 LTS (64 bit)
Virtual Box	Oracle cross-platform virtualization application
WANem	Wide Area Network Emulator version 2.3
Streaming Server	Ngnix web server with an additional module for RTMP
OS Client	Ubuntu 16.04.4 LTS (64 bit)
Browser	Opera
Web Interface	Custom Interface
Video Player	JWPLAYER
Streaming technologies	HLS, RTMP, MPEG-DASH

The following limitations were used in the experimental platform, drawing on the results of our previous research [4] and in line with new trends:

- The bandwidth of the path was not limited.
- We changed streaming technologies to HLS, RTMP, and MPEG-DASH, which are now more popular than others [23].

- The video clip was chosen with a resolution of 1920 x 1080 pixels for the next step in our research [4]; detail parameters are presented in Table 3.

Server, WANem, and client were virtualized on VirtualBox. They were all connected to the virtual switch. WANem was used to emulate internet cloud behaviour. Traffic between client and server was routed through the WANem virtual appliance.

Table 3. Detailed parameters of the video file

Name	**Description**
Name	Big Buck Bunny
Resolution	1920*1080 FHD
Video Codec	H.264
Frame per second	24 fps
Video Bitrate	9283 kbps
Audio Codec	MPEG4-AAC
Audio bitrate	437 kbps

Results

A. General research results

The first study was conducted for the single factor effect of packet loss and delays on the start of re-buffering using the technique proven in previous research [4]. The new streaming technologies, HLS and MPEG-DASH, were added to the experimental platform, reflecting current trends in video streaming [23]. From the results of the study, we identified the beginning of re-buffering in the streaming protocols RTMP, HLS, and MPEG-DASH, as presented in Table 4.

Table 4. Streaming technology comparison (influence of a single parameter on the start of re-buffering)

	Streaming Technologies			**ITU-T Y.1541 objective**
	RTMP	HLS	MPEG-DASH	
Packet loss, %	4.2	4.7	5.3	0.1
Delay, ms	287	180	242	1000

We analyzed the results obtained in terms of the current requirements to ensure QoS for different traffic classes under ITU-T Recommendation Y.1541 (12/2011) [11]. Streaming video corresponds to class 4, for this class the maximum total allowable delay is 1,000 ms and packet loss must be less than 0.1 %. The means that delays are a more significant factor for re-buffering than packet loss. Moreover, exceeding the limits of delays indicated in Table 4 is guaranteed to result in re-buffering even if packet loss is zero. In the Recommendations, rather big delays are allowed; in reality, however, delays of 180-287 ms cause re-buffering. The main reason for this effect is that in the Recommendations for streaming video, transmission was supposed to use the UDP/IP stack; in OTT technology the TCP/IP stack is used. While using TCP, retransmission can occur increasing the delay of a single packet significantly and causing the buffer to empty and leading to re-buffering. This means that with mean delay values of 200-300 ms re-buffering can occur.

In modern computer networks, packet loss is less than 0.1 % and delays are less than 100 ms so that the thresholds for re-buffering decrease significantly.

Research results on the influence of two factors (IPTD and IPLR) on re-buffering are presented in Table 5. During research, we fixed the delay and changed the packet loss until re-buffering started.

Table 5. Streaming technology comparison (influence of the parameters in combination on re-buffering)

Delay, ms	**Packet loss, %**		
	RTMP	HLS	MPEG-DASH
4	0.37	0.65	0.82
5	0.16	0.25	0.35
7	0.14	0.21	0.28
9	0.08	0.1	0.17

The results are in the range of QoS parameters of real modern data networks; QoS parameters should be monitored and controlled to enable video streaming transmission.

As can be seen in Fig. 6, the best protocol for streaming video is MPEG-DASH, second is HLS, while the worst is RTMP. The slight difference between the streaming protocols can be explained by peculiarities in the implementation of the algorithm of each specific protocol and the difference in header size and service messages.

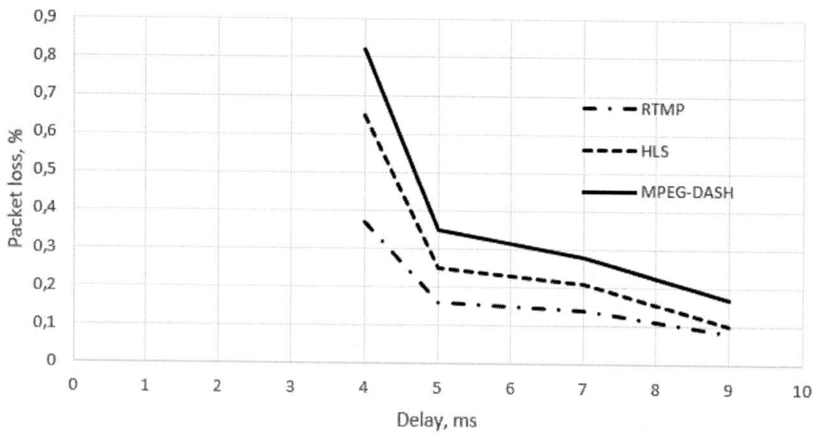

Fig. 6. Thresholds of re-buffering for HLS, RTMP and MPEG-DASH

The other meaningful factor that influences the start of re-buffering, is the effective bandwidth of the TCP connection. In Annex 9 of ITU-T Y.1541 (12/2011) [11], the calculation of the effective TCP session rate is discussed in detail. We use the formula for calculating the effective throughput of a TCP session presented in this Recommendation in the fifth section of this paper. We obtained a theoretical maximum of TCP throughput from packet loss and delay in using it. This will be the upper limit of video AVG (average) bitrate.

B. Re-buffering time-frame

During our research, we found that re-buffering effects occur only at a specific time-frame of video playback. We took three different clips with resolution at 1920 x 1080 FHD and analyzed their bitrate using the Bitrate Viewer [24] software:

- Video 1: Big Buck Bunny (Fig. 7), used in current research.
- Video 2: Sport (Fig. 8) used in previous research [4, 8-10].
- Video 3: Nature (Fig. 9) used in previous research [4, 8-10].

The results of this analysis showed that each tested video has a specific bitrate, but they all have variable bitrate (VBR). VBR is used by default for most video on the Internet because it gives much greater quality at the same AVG bitrates. It is an encoding method that enables the bitrate of a video to dynamically increase or decrease according to the complexity of video fragments. For more complex fragments, the bitrate will be higher while for less complex fragments the bitrate will be lower. Constant bitrate (CBR), in contrast, utilizes a set bitrate over the entire video clip leading to image quality being degraded during complex video segments.

Analysis of this video is presented in Table 6. The videos have nearly the same AVG bitrate, but extremely different MAX and MIN values at specific time-frames of playback. Based on this analysis of video bitrate, we can predict for which time-frames of video playback consumers will frequently see re-buffering. For Video 1, re-buffering can occur at the beginning, in the middle, and at the end of the video. For Video 2, re-buffering is likely to occur at the beginning and after the middle of the video. And in Video 3, re-buffering could occur at any time, because the compensation buffer increases the possibility of re-buffering throughout.

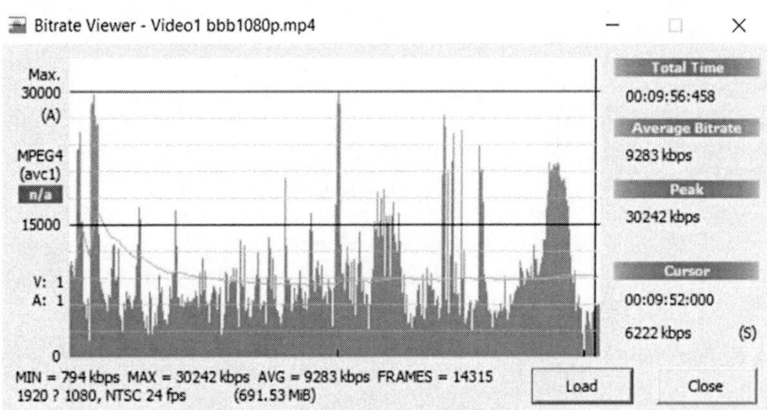

Fig. 7. Bitrate of Video 1

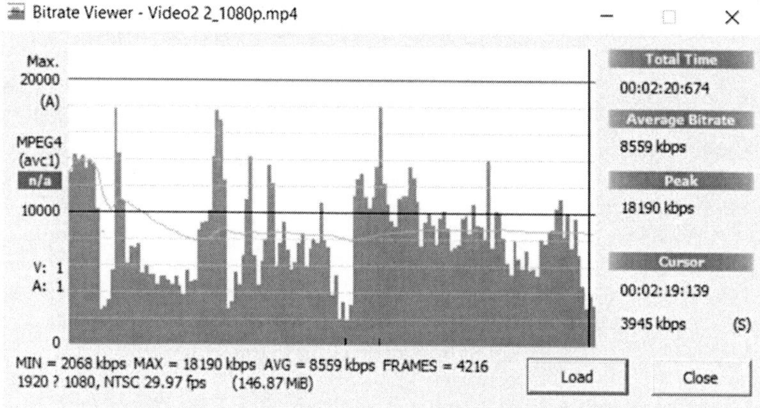

Fig. 8. Bitrate of Video 2

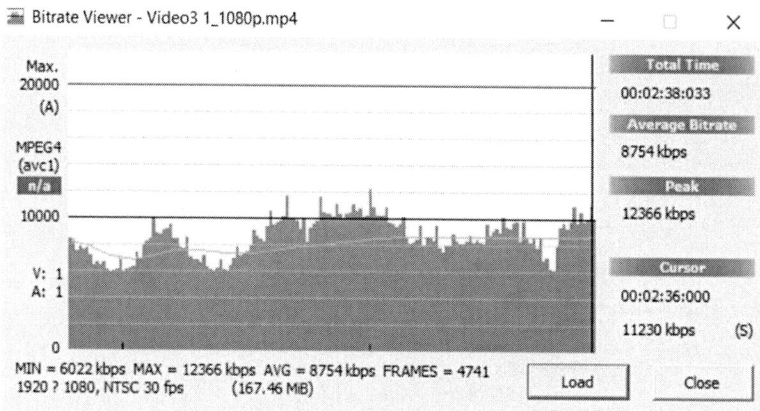

Fig. 9. Bitrate of Video 3

Table 6. Parameters of video bitrate for videos 1, 2, & 3

Parameter	Video 1	Video 2	Video 3
AVG bitrate, kbps	9283	8559	8754
Min bitrate, kbps	794	2068	6022
Max bitrate, kbps	30242	18190	12366

Research into QoS Parameters Thresholds

Analysis of the results showed that TCP throughput has a significant influence on re-buffering. Using formula [11] to calculate the effective speed of a TCP session, we can estimate the opportunity to transfer streaming video over the network with defined QoS parameters and without re-buffering.

It is not straightforward to compare the AVG of streaming video and throughput of a TCP session. It is necessary to consider additional headers and service packets of these streaming technologies. This can lead to a significant reduction in throughput during video transmission.

For the creation of a graphic presentation of TCP throughput as a function of packet loss and delay, we can use the formula of the simple model of TCP throughput from [11]:

$$B(p) \approx \min\left(\frac{W_{max}}{RTT}, \frac{1}{RTT\times\sqrt{\frac{2bp}{3}}+T_0 min(1{,}3\times\sqrt{\frac{3bp}{8}})\times p\times(1+32\times p^2)}\right) \quad (1)$$

where W_{max} is the maximum window buffer size of a receiver (packets); RTT is the round trip time (sec.); b is the number of packets that are acknowledged by a received ACK; p is the probability that a packet is lost; and T_0 is the time-out for re-transmitting an unacknowledged (lost) packet (sec.).

We can use the following simplification for formula 1, because packet loss is less than 0.1 % in modern networks:

$$B(p) \approx \frac{1}{RTT\times\sqrt{\frac{2bp}{3}}+T_0 min(1{,}3\times\sqrt{\frac{3bp}{8}})\times p} \quad (2)$$

From formula (2) we can use the formula for TCP throughput:

$$T = B(p) \times MSS = \frac{MSS}{RTT\times\sqrt{\frac{2bp}{3}}+T_0 min(1{,}3\times\sqrt{\frac{3bp}{8}})\times p} \quad (3)$$

where MSS is the maximum segment size.

Table 7 presents recommendations from Google [25] for uploading a video to YouTube. If we compare the recommendations in Table 7 and a 3D chart of the maximum TCP throughput in Fig. 10, we can see that it is possible to watch a video at 4K resolution with current standard network

QoS parameters if we use a stable cable connection with a bandwidth of more than 30 Mbps. However, for an 8K video, there is a high probability that we will see re-buffering. In modern telecommunication networks, it is very difficult to ensure a delay of less than 15 ms and a packet loss of less than 0.03 % continuously. As such, these QoS parameters may be guaranteed only by a CDN provider that has servers in each city. This restricts the possibility of expanding 4K and 8K services from OTT providers.

To watch 8K UHD video from an OTT service provider, the client must have a stable internet connection with a minimum of 50 Mbps speed and additional delay and packet loss on the client side.

Table 7. Video bitrates for different resolutions

Name	Resolution	Bitrate, Mbps
SD	720 × 576	2.5
HD	1280 × 720	6.5
FHD	1920 × 1080	10
4K UHD	3840 × 2160	25
8K UHD	7680 × 4320	40

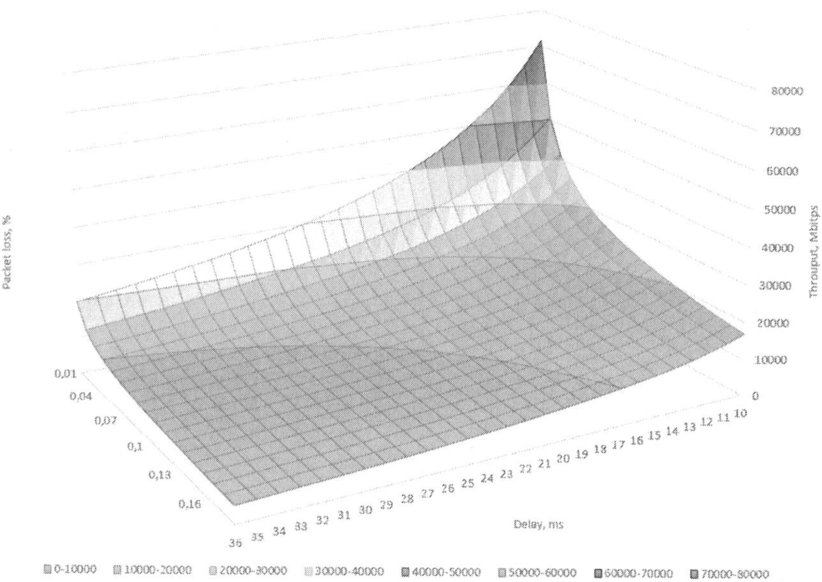

Fig. 10. Maximum TCP throughput with packet loss and delay

TCP Alternatives

The effective throughput of a TCP session depends on delay and packet loss and this is the primary cause of re-buffering. The calculated effective TCP throughput for a given delay and packet loss allows us to estimate the possibility of delivering a video stream at a definite bitrate. As such, TCP session throughput functions as a bottleneck for UHD video streaming via OTT.

We need an alternative transport protocol to overcome the limitations of TCP. This protocol must be as reliable as TCP and have throughput that does not depend on delay. In Table 8, we present some transport protocols [26].

Table 8. Comparison of transport layer protocols

Feature	UDP	TCP	RUDP	SCTP	MPTCP	QUIC
Connection-oriented	No	Yes	Yes	Yes	Yes	Yes
Reliable transport	No	Yes	Yes	Yes	Yes	Yes
Ordered delivery	No	Yes	Yes	Yes/ No	Yes	No
Multihoming	No	No	No	Yes	Yes	Yes

Connection-oriented protocols set up a connection before data delivery. For example, TCP uses a three-way handshake to set up a connection.

Reliable transport guarantees packet delivery over the network.

Order delivery ensures that packets are delivered in the same order that they were sent.

Multi-homing gives clients and servers the possibility of setting up multiple connections using different addresses. This may increase reliability/performance.

A. RUDP

The first draft of the Reliable User Datagram Protocol (RUDP) was published in 1999 [27], but this protocol has not yet been ratified. Cisco and Microsoft have both created versions of this protocol, but they are not compatible with each other and differ from the IETF draft. As a result, it is not currently possible to use this protocol as an alternative to TCP.

B. SCTP

Stream Control Transmission Protocol (SCTP) was ratified in October 2000 [28, 29]. SCTP was designed to transport public switched telephone network (PSTN) signalling messages over IP networks, but it can be used for other purposes.

Advantages of SCTP over TCP:

- SCTP is message-oriented;
- SCTP supports multiple streams;
- SCTP supplies multi-homing;
- SCTP is more resistant to man-in-the-middle and DoS attacks.

There are some reasons why we do not use it in this way:

- The main reason is that SCTP came after TCP and failed to push it aside;
- Developers do not really know about the benefits of SCTP;
- We must change the transport level of the TCP/IP stack on the host;
- Applications have to be modified to manage the SCTP's multiple streams;
- NAT is another problem as most routers and firewalls support NAT translation only for UDP and TCP traffic.

It is very difficult to implement a new protocol. For example, IPv6 has been in the process of implementation for the past 10 years, but it is still not yet used everywhere. Unfortunately, it is very difficult to implement the new SCTP transport protocol, because this will require significant changes to the Internet. This will be very difficult to do, as the widely used TCP and UDP meet most current needs.

C. MPTCP

Multipath TCP (MPTCP) was published as an experimental RFC in January 2013 [30, 31]. This is an extension of the TCP protocol. The primary purpose is to have the capability of simultaneously setting up TCP connections by different paths. MPTCP will increase the throughput and resilience of a single TCP session by using multiple interfaces.

The IETF working group dictates two key restrictions on Multipath TCP design in RFC 6824 [30]:

- It must be backwards-compatible with current regular TCP to increase its chances of deployment.
- It can be assumed that one or both hosts are multi-homed and multi-addressed.

The new version of MPTCP is still in draft and support for its development by large companies has almost stopped. For example, Google stopped developing this project and focused on the development of an alternative QUIC Protocol. As such, the implementation of MPTCP has ground to a halt.

D. QUIC

The first IETF draft of the Quick UDP Internet Connection (QUIC) was published in 2016 [32]. QUIC provides the following functionality:

- UDP based;
- Stream multiplexing;
- Flow control;
- Low connection latency;
- Connection rehousing;
- Supports NAT rebinding;
- Security.

QUIC is currently the best alternative to TCP and other transport protocols:

- QUIC uses UDP as an underlying protocol, so we do not need to change the client's operating system and think about how to resolve problems with NAT;
- Google and Facebook implement QUIC in their services; Microsoft has also started to develop QUIC;
- QUIC has minimal connection latency, helping to minimize the delayed start problem;
- The Multipath QUIC extension [33] is a good alternative to the MPTCP protocol.

Unfortunately, QUIC is only at a developmental stage right now (although it is currently undergoing rapid development). The lack of a final RFC version and many implementations by Google does not yet allow the widespread implementation of this protocol.

Conclusion

As OTT becomes more and more popular among national and international providers, the quality of experience for streaming technologies becomes more and more important.

In this work, we have confirmed our previous results suggesting that the combination of delay and packet loss has significant influence on re-buffering in the OTT technologies: HLS, RTMP, and MPEG-DASH.

The primary cause of re-buffering is that the effective throughput of a TCP session is dependent on delay and packet loss. The calculated effective TCP throughput for a given delay and packet loss allows us to estimate the possibility of delivering video streams at definite bitrates. As such, TCP session throughput functions as a bottleneck for UHD video streaming via OTT services.

Our research has shown that using only one TCP session for transport is a significant restriction. Any video streaming protocol will be limited by TCP session throughput. This important limitation has to be taken into consideration when design network protocols for new NGN and FGN networks.

Future Work

Further study should be devoted to research into how different TCP algorithms influence re-buffering. A lot of combinations of algorithms should be investigated at both the client and the server side.

We are planning to extend our experimental platform for research into acceptable QoS parameters for 4K and 8K UHD video.

The main goal of our future work will be to create a new method for evaluating the thresholds of different combinations of QoS parameters for different streaming technologies and to find a correlation between AVG video bitrate and TCP throughput.

Also, we intend to undertake research into the QUIC protocol as a primary alternative for TCP in more detail, in terms of checking error correction, especially when the bit error level is high.

References

[1] “Cisco Visual Networking Index: Forecast and Trends, 2017–2022,” Cisco Systems Inc., https://www.cisco.com/c/en/us/solutions/collateral/service-provider/visual-networking-index-vni/white-paper-c11-741490.html.
[2] “The State of Online Video,” Limelight Networks, https://www.limelight.com/resources/white-paper/state-of-online-video-2017.
[3] “The Future of OTT Streaming,” TV Technology, https://www.tvtechnology.com/opinions/the-future-of-ott-streaming.
[4] V. U. Deart and A. A. Nesterkin, “A Study of Influencing Factors on the Quality of Experience Perception in Video Streaming,” Proceedings of the 18th Conference of Open Innovations Association FRUCT, Apr. 2016, pp. 437-442.
[5] “OTT viewers now care more about QoE than the choice of content,” Videonet White Papers, http://www.v-net.tv/ott-viewers-now-care-more-about-qoe-than-choice-of-content-says-conviva.
[6] “State of Streaming report,” IBM, http://info.video.ibm.com/State-of-Streaming-2017-Report.html.
[7] “Bit rate and business model,” Akamai Technologies, https://www.akamai.com/us/en/multimedia/documents/white-paper/bit-rate-and-business-model.pdf.
[8] V. Deart, I. Kozhuhov “Research of the QoS parameters defining QoE for transmitting streaming video over the Internet," T-Comm, no. 7, 2013, pp. 28-31.
[9] V. Deart, I. Kozhuhov, A.V. Pilugin “Experimental Platform for QoE research of streaming video services”, T-Comm, no. 7, 2013, pp. 32-35.
[10] V. Deart, I. Kozhuhov, Y. Milicin, A. Pilugin “Analysis of experimental data on subjective assessment of QoE when transmitting video content using OTT”, T-Comm, no.8, 2014, pp. 37-42.
[11] “Recommendation ITU-T Y.1541 (12/2011) Network performance objectives for IP-based services,” ITU-T, http://www.itu.int/rec/T-REC-Y.1541.
[12] “Real-Time Messaging Protocol (RTMP) specification,” Adobe, http://www.adobe.com/devnet/rtmp.html.
[13] “Recommendation ITU-T G.1080 (12/2008) Quality of experience requirements for IPTV,” ITU-T, http://www.itu.int/rec/T-REC-G.1080/en.
[14] “BOLA: Near-optimal bitrate adaptation for online videos,” IEEE, https://ieeexplorc.ieee.org/document/7524428.

[15] "Study of user QoE improvement for dynamic adaptive streaming over HTTP (MPEG-DASH)," IEEE, https://ieeexplore.ieee.org/document/7876191.
[16] A. Zabrovskiy, E. Petrov, E. Kuzmin and C. Timmerer "Automated Performance Evaluation of Adaptive HTML5 Player Deployments," Proceedings of the 21st Conference of Open Innovations Association FRUCT, Nov. 2017, pp. 532-536.
[17] "HTTP Live Streaming," Apple, https://developer.apple.com/streaming.
[18] "MPEG-DASH," Moving Picture Experts Group, https://mpeg.chiariglione.org/standards/mpeg-dash.
[19] "Internet Connection Speed Recommendations," Netflix, https://help.netflix.com/en/node/306.
[20] "Nginx HTTP server features," Nginx official website, https://nginx.org.
[21] "NGINX-based Media Streaming Server," GitHub, github.com/arut/nginx-rtmp-module.
[22] "Wide Area Network Emulator," SourceForge, https://sourceforge.net/projects/wanem.
[23] "2018 Global Media Formats Report," Encoding.com, https://www.encoding.com/resources/#white-papers
[24] "Bitrate Viewer 2.3," VideoHelp, https://www.videohelp.com/software/Bitrate-Viewer-2.
[25] "Live encoder settings, bitrates, and resolutions," Google, https://support.google.com/youtube/answer/2853702?hl=en.
[26] "Transport layer," Wikipedia, https://en.wikipedia.org/wiki/Transport_layer#Comparison_of_transport_layer_protocols
[27] "Reliable UDP Protocol," IETF, https://tools.ietf.org/html/draft-ietf-sigtran-reliable-udp-00
[28] "Stream Control Transmission Protocol," IETF, https://tools.ietf.org/html/rfc2960
[29] "What About Stream Control Transmission Protocol (SCTP)?," IDG Communications, Inc., https://www.networkworld.com/article/2222277/what-about-stream-control-transmission-protocol--sctp--.html
[30] "TCP Extensions for Multipath Operation with Multiple Addresses," IETF, https://tools.ietf.org/html/rfc6824
[31] "TCP Extensions for Multipath Operation with Multiple Addresses draft-ietf-mptcp-rfc6824bis-16," IETF, https://tools.ietf.org/html/draft-ietf-mptcp-rfc6824bis-16
[32] "QUIC: A UDP-Based Multiplexed and Secure Transport," IETF, https://datatracker.ietf.org/doc/draft-ietf-quic-transport/
[33] "Multipath Extensions for QUIC (MP-QUIC) draft-deconinck-quic-multipath-02," IETF, https://tools.ietf.org/html/draft-deconinck-quic-multipath-02

CHAPTER 1.2

ENSURING THE RELIABILITY OF SOFTWARE DEFINED NETWORKS

VICTOR NETES, MARGARITA KUSAKINA

Abstract. The software-defined network (SDN) is a developing technology in telecommunications. Along with a number of other advantages, its usage will automate the management and administration of network equipment and greatly accelerate the organization of new services for users. A new control element, the "brain" of the SDN, is a controller that performs the functions of controlling physical devices (switches) while at the same time interacting with the application layer. One of the most important characteristics of any communication network is reliability; reliability requirements are increasing with the expansion and intensification of usage of information and communication systems. This paper discusses the main aspects of SDN reliability. As the controller is a key element of centralized network control, its failure or a loss of connection with switches will stop the normal functioning of the network. Therefore, it is important to ensure the high reliability of the controller and its connections to network elements. With the example of a typical SDN, several options for placing and linking controllers, and the redundancy of their connections, are analyzed. Calculations show the need for redundancy in controllers and their connections to all nodes in the network.

Introduction

In recent years, one of the most widely discussed topics in telecommunications is that of the software-defined network(ing) (SDN) [1]. The SDN concept has emerged in response to the growth of virtualization, mobility, and the Internet of things, etc. It is a new evolutionary concept in network architecture, which sees the separation of the control plane from the data plane. In SDN, the data plane forwards network traffic according to instructions from the control plane.

The main advantages of SDN are as follows. Firstly, it centralizes management of networking devices and has improvements for end users. Secondly, SDNs are more flexible and scalable compared to traditional networks. Lastly, it offers geographical independence in terms of the placement of the platform.

Standardization in the field of SDNs is carried out by a number of international organizations, in particular the ITU-T Recommendations Y.33xx are devoted to this topic. ITU-T considers the SDN concept to be an important shift in network technology that will enable network operators to create and manage new virtualized resources and networks without deploying new hardware technologies.

Many challenges still remain to be solved; these include problems of scalability, virtualization, communication consistency, controller placement, and so on. Reliability is a major challenge in SDNs. As stated in [2], in large-scale networks, reliability is a particularly important issue. Characteristic of the logically centralized control in SDN, the controller tends to become a single point of failure. Therefore, it is necessary to take measures to ensure the reliability of such new technical solutions that are at least as good, or even better, than those used previously. SDN is considered to be one of the most important technologies for building the network infrastructure of the digital economy [3]. However, the digital economy cannot be based on unreliable infrastructure.

The aim of this paper is to discuss the main aspects of ensuring reliability in SDNs. In particular, we analyze the controller placement problem, controller redundancy, and redundancy in connections between controllers and network elements.

The rest of the paper is organized as follows. The following section is devoted to the evolution of networks and the emergence and development of SDN; it also describes the architecture of the new paradigm, its application in future and 5G networks, and its development in Russia. The third section presents an overview of reliability and two important aspects in communications networks. The fourth section describes the problem of controller placement in terms of reliability and presents possible solutions. The fifth section considers a typical SDN and a few redundancy options for controllers and connections. The sixth section discusses the problem of calculating network reliability, taking into account the specificity of the SDN. The concluding section gives the main findings and offers some directions for future work.

The Concept and Use of SDN

A. Brief history of its evolution

The history of SDN principles can be traced back to the separation of the control and data plane, which was first used in the public switched telephone network as a way of simplifying provisioning and management well before its use in data networks [4]. The use of open source software in split control/data plane architectures traces its roots to the Ethane project at Stanford University's Department of Computer Sciences. Ethane's simple switch design led to the creation of the OpenFlow protocol. An application program interface (API) for OpenFlow and an operating system for networks (NOX) were created in 2008.

Work on OpenFlow has continued at Stanford and has included the creation of testbeds to evaluate the use of the protocol in a single campus network, as well as across a WAN as a backbone architecture for connecting multiple campuses. In academic settings, a few research and production networks based on OpenFlow switches have been made by NEC and Hewlett-Packard; as well as ones based on Quanta Computer whiteboxes starting from 2009.

In 2011, the Open Networking Foundation was founded to promote SDN and OpenFlow.

B. The definition and architecture of SDN

ITU-T Recommendation Y.3300 [2] gives the following definition for the term software-defined networking: a set of techniques that enables one to directly program, orchestrate, control and manage network resources, which facilitates the design, delivery and operation of network services in a dynamic and scalable manner.

It is known that any network has both control and data planes. The control plane is generally considered to be where a router or switch makes its decisions. In this case, it is software based and uses the CPU, rather than any specialised hardware. The data plane (or forwarding plane) is the high speed path through the router/switch. Packets that pass through the device use the data plane.

In traditional IP networks, the control and data planes are tightly coupled, embedded in the same networking devices, and the whole structure is highly decentralized (Fig. 1).

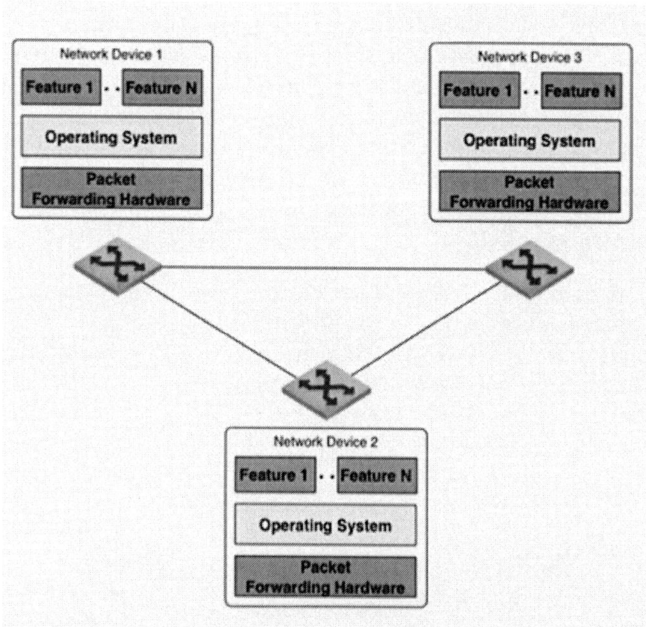

Fig. 1. Architecture of traditional IP networks

In SDN, the control and data planes are separated (divided between controllers and network devices), and the whole structure is centralized. The architecture of an SDN consists of three layers [2], as depicted in Fig. 2. The interaction between them is realized through APIs. These are presented in more detail below.

1) Application layer: The application layer is where SDN applications specify network services or business applications by defining a service-aware behavior of network resources in a programmatic manner. These applications interact with the SDN control layer via APIs (northbound interfaces) in order for the SDN control layer to automatically customize the behavior and properties of network resources. The programming of an SDN application makes use of the abstracted view of network resources provided by the SDN control layer and by means of information and data models exposed via the northbound interface.

2) SDN control layer: This layer provides a means to dynamically and deterministically control the behavior of network resources (such as data transport and processing), as instructed by the application layer. SDN applications specify how network resources should be controlled and allocated, by interacting with the SDN control layer via the northbound interface. The control signaling from the SDN control layer to the network resources is then delivered via resource-control interfaces (southbound interfaces). The configuration and/or properties exposed to SDN applications are abstracted by means of information and data models. The level of abstraction varies according to the applications and the nature of the services to be delivered.

3) Resource layer: The resource layer is where the network elements perform the transport and processing of data packets according to the decisions made by the SDN control layer, which have been forwarded to the resource layer via a southbound interface.

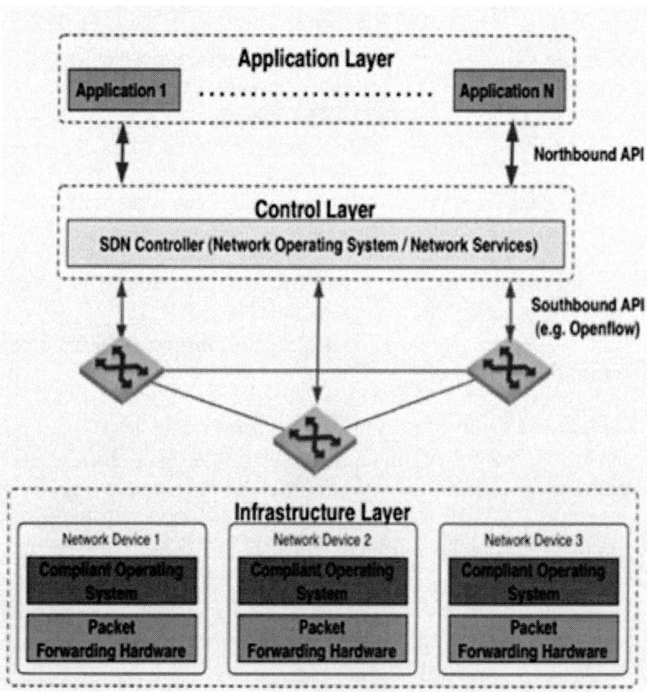

Fig. 2. Architecture of an SDN

C. Use in future networks and 5G

Since 2011, the ITU-T has adopted a number of Recommendations (Y.30xx) on so-called future networks (FNs), as the generation of networks following next generation networks (NGNs). There are 12 design goals for FNs, being high-level capabilities and characteristics that should be supported [5]. Two of these goals are reliability and security. In particular, it is recommended that FNs are designed, operated, and developed with reliability and resilience in mind and taking into account challenging conditions. Software-defined networking is a core technology for FNs [2].

Another area of telecommunications being actively developed is the fifth generation of mobile networks (5G). In the ITU, the 5G standard is named IMT-2020. The letters in this designation stand for international mobile telecommunications and the numbers indicate the year that it should be completed (for the third generation, a similar designation was used: IMT-2000). ITU-T Recommendation Y.31xx is devoted to this topic.

One of the requirements for the IMT-2020 network is that it should be designed with fault tolerance and operate reliably [6]. At the same time, it is stipulated that reliable and stable operation is especially important in the event of network overloads and emergencies. In addition, it is stated that the reliability and resiliency of the network should not be put at risk by any software or hardware upgrade.

The software is mentioned here not by chance, but because one of the characteristic features of IMT-2020 is network "softwarization" [7]. This means an overall approach is taken to designing, implementing, deploying, managing, and maintaining network equipment and/or network components through software programming [8]. Basic technologies for network softwarization include: software-defined networking, network function virtualization (NFV), and cloud computing [7]. At the 2014 Mobile World Congress, there was a clear consensus that future networks will revolve around virtualization and SDNs [9]. As such, SDN reliability will have a direct impact on the reliability of 5G networks.

D. Current activities

The testing and implementation of SDN solutions is being carried out by a number of American and European telecom operators, including:

AT&T, BT, DT, EE, Telefonica, Verizon, Vodafone, etc. (a brief overview of their activities can be found in [3]).

The largest Russian telecom operator, Rostelecom, is also testing SDN solutions in its laboratory. Studies have highlighted a number of problems and shortcomings preventing the application of these solutions in real networks [3], [10]. A large part of these shortcomings are directly related to reliability, including:

- A long convergence time in the case of connection failure (of a few seconds);
- losses of traffic when controllers are recovering from such a failure;
- the inability to detect some faults in channels.

The development of solutions for the SDN is also actively being conducted in the Applied Research Center for Computer Networks (ARCCN) [11]. This is the first Russian research center created to develop technologies and products for new generation computer networks. The ARCCN startup, RunSDN, has a complex all-in-one solution for organizing the transport SDN network; it is the only Russian company to have passed the threshold for participation in the SDN & NFV Solutions Showcase at ONS 2017 [11].

Equipment and System Reliability

A. Basic concepts and measures

In this paper, we are discussing reliability. Strictly speaking it would be more correct to use the term dependability. The fundamental international standard in this area can be found in [12]. Under the agreement between the International Electrotechnical Commission (IEC), the International Organization for Standardization (ISO), and the International Telecommunication Union (ITU), constituting the World Standards Cooperation, it is the IEC that plays the leading role in standardization in this area.

In [12], dependability is defined as an item's ability to perform as and when required. Dependability is used as a collective term covering the time-related quality characteristics of an item; it includes the following attributes: availability, reliability, recoverability, maintainability, and maintenance support.

An item within this standard may be an individual part, a component, a device, a functional unit, a piece of equipment, a subsystem, or a system. The item may consist of hardware, software, people, or any combination thereof. In our consideration, this term can be used both for the network as a whole, as well as for its elements: nodes, links, and terminals, etc.

In turn, availability is the ability of a subject or component to be in a state to perform as required. This depends upon a combination of the reliability, recoverability, and maintainability of the item, and its maintenance support performance. Reliability is the ability to perform as required, without failure, for a given time interval and under given conditions.

The term dependability is used in IEC standards devoted to communication networks [13], [14] and in ITU-T Recommendation E.862 [15]. Unfortunately, quite often the term reliability is used, in a broad sense, instead of the term dependability [16]. As such, in many other ITU-T Recommendations, in numerous publications, in the names of many journals, and at conferences, etc. there is a certain lack of rigor in its usage. To maintain consistency with these other sources, we will also write about reliability, although it would be more correct to talk about dependability.

The terms reliability and availability are used for quantitative measures of appropriate attributes [12]:

- Reliability is the probability of performing as required for a given time interval and under given conditions.
- Availability is the probability that an item is in a state to perform as required at a given instant (instantaneous/point availability), or the limit, if it exists, of the instantaneous availability when the time tends towards infinity (steady state/asymptotic availability). The most commonly used form is steady state availability, which is often merely called availability.

Both these measures are probabilistic in nature and can be expressed as fractions or percentages; however, the latter form is unsuitable for use in calculations. The values 0.99 (99 %) and 0.999 (99.9 %), etc. are often called “two nines”, “three nines”, etc.

Additionally, availability can be expressed in terms of downtime per year, or for another time period. For example, service level agreements often refer to monthly downtime in order to calculate service credits to match monthly billing cycles. The downtime for the period T corresponding to the availability A is calculated by the formula

$$D_T = (1 - A)T. (1)$$

B. Two aspects of reliability

Usually, two aspects of reliability are considered for communication networks: equipment and system (or structural) reliability [17], [18]. The first one depicts the reliability of equipment, which is included in network nodes. The system aspect, also called the structural aspect, shows the functioning of the network as a whole, depending on the performance or failure of nodes and links.

C. Five nines as a target

Currently, the generally accepted reliability and availability requirement for carrier-class network equipment and critical applications is a value of 0.99999 (see [18], [19], [20], [21]). As an aside, one of the companies incorporated in the USA during the immense telecom industry turmoil of 2002, was named the Five Nines Networks [22]. The average downtime corresponding to this value of availability according to (1) is 5.256 minutes per year.

D. Equipment reliability

A controller is a new equipment type in an SDN. It is the key element of the network. Sometimes it is termed the "brains" of the network and, as such, it is necessary to ensure a high level of reliability.

As has already been mentioned, there is a danger that a controller will be a single point of failure [2]. As noted in [2], the SDN controller can be replicated to improve reliability. In addition, the use of only one controller is not desirable, keeping in mind the need to ensure network survivability, i.e. the ability for the network to continue to function during and after a natural or man-made disturbance. All this suggests the need for redundancy for controllers with geographical separation of their locations (geo-redundancy). A similar situation arose previously in NGNs in relation to their key element: the softswitch.

Two approaches are possible when arranging the controller. Firstly, it can be made as a traditional standalone piece of specialized equipment; secondly, it is possible to virtualize its functionality by placing the controller in the cloud, i.e. in a data center, resulting in the symbiosis of SDN technology and network function virtualization (NFV).

If we consider these two approaches from the point of view of reliability, if the controller is implemented in a traditional way and placed in a node, it is closer to the elements of the network, giving lower latency and higher reliability of communication between the controller and the controlled network elements. If it is implemented in a data center, its reliability may be higher, but there is now a greater distance between the controller and the network elements. This results not only in higher latency, but also less reliable communication between the controller and the network elements.

This circumstance is typical of cloud computing in general. It is no coincidence that [21] is entitled "How Low-cost Telecom Killed Five 9s in Cloud Computing". However, leading providers and vendors have set the task of achieving "five nine" reliability in a telco-grade cloud [23].

General considerations about the reliability and availability of cloud services with calculations for typical situations and recommendations are presented in [24] and [25]. Among the main findings of these papers are the following: the availability of cloud services should be estimated on an end-to-end basis, taking into account all components of a work path; achieving high availability requires redundancy for data centers and network connections.

E. System reliability

New tasks related to system reliability in SDNs include the selection of the number and locations of controllers; the distribution of controlled network elements between them; and the arrangement of communication between controllers and the network, etc.

The SDN provides new opportunities to increase reliability through self-recovery and the automatic redistribution of traffic flows (re-routing). These are provided by the OpenFlow protocol, which finds a new route when necessary; however, it is undesirable to overload the backup path links. Rational organization of such re-routing requires the solution of the above-mentioned tasks.

SDN Controller Placement

A. General overview

The issue of controller placement has been investigated in a number of papers [26], [27], [28], [29], [30]. Most of these focused on performance and latency. In particular, [27] considers the problem and offers solutions to the following: finding the minimum number of controllers in the network; the choice of locations of controllers among candidate nodes; the distribution of controlled network elements between controllers; and the reassignment of controllers in case of failure with minimal deterioration. The limitations on packet transmission latency between the controller and the equipment and between controllers, as well as the limitations due to the need for load balancing between controllers, all need to be taken into account. In the case of controller failure, for backup the above conditions should be maintained with minimal losses.

Work making direct reference to reliability can be divided into two groups. In some, the task was to optimize (maximize) reliability. In others, the goal was to minimize cost, and reliability was taken as a constraint. This is similar to the two main optimization tasks of optimal redundancy [31].

In [30], two reliability optimization problems are considered. For the case of the shortest path between the controller and the switches, a clustering-based global optimization algorithm and suboptimal algorithm based on a greedy algorithm are proposed. The optimization problem is how to realize controller placement in relation to the switches to maximize the average reliability of the network. For the case of multi-paths between the controller and the switches, in order to simplify computational complexity, the heuristic reliability factor is introduced; then, the modified clustering-based global optimization algorithm and the suboptimal algorithm are proposed. Unfortunately, a number of typographic errors make it difficult to fully understand this paper.

In [29], a heuristic algorithm that computes placement with (at least) required reliability is proposed. The objective is to minimize the total cost of deploying controllers and connecting nodes to them. The requirement for the reliability of communication between controllers and nodes (southbound reliability) is set to 0.99999 (five nines). In order to simplify the computational complexity, instead of an exact value for southbound reliability, its lower bound is taken based on a set of disjointed paths. This is a well-known bound in reliability theory, termed in [31] the lower Litvak-Ushakov bound. It is obtained by parallel connection of disjointed minimal paths, each of which is a serial connection of its elements.

B. Latency metrics and their modification for reliability optimization

When solving the problem of controller placement, various metrics can be used. The most popular metrics are related to latency [28]:

- the minimum—average latency,
- the minimum—worst-case latency.

They are respectively defined by the formulas:

$$L_{avg}(S) = \frac{1}{n}\sum_{v\in V} \min_{s\in S} d(v,s)\,, (2)$$

$$L_{wc}(S) = \max_{v\in V} \min_{s\in S} d(v,s)\,, (3)$$

where S is the set of nodes in which the controllers are placed; V is the set of nodes in which network equipment is placed; n is the number of nodes in the set V; and d is the latency between the indicated nodes.

The optimization problems with metrics (2) and (3) are reduced to the well-known graph theory problems of finding medians and centers, respectively [32] and [33]. They are used in various applications. For example, the median might be a good place to locate a mall: the average driving distance is minimized and the center would be a good place for the location of an emergency facility, as the response time must be minimized in the worst case scenario. The problems of placing several medians and centers are also known. The *m*-median problem is how to locate *m* facilities on a network so as to minimize the average distance from one of the demand points to one of the *m* facilities. The *m*-center problem is how to locate *m* facilities on a network so as to minimize the largest distance from a demand point to its nearest facility. In the situation considered, the facilities are SDN-controllers and the demand points are switches at network nodes.

The latency in (2) and (3) is the sum of the delays on all links in the minimal path between the considered nodes:

$$d(v,s) = \sum_{(i,j)\in P} d_{ij}, (4)$$

where d_{ij} is the delay in the link between nodes i and j; P is the set of links that make up the path between nodes v and s.

For these tasks, we consider a weighted graph in which "weights" are assigned to the arcs representing their lengths or delays. However, we can apply these results to reliability. The considered approach can be used to optimize reliability using the following idea [33]. In shortest path problems, the sum of the weights of the arcs forming the path are taken as the path length. Consider now the case where the weight of an arc represents its reliability. The reliability of the path P is calculated by the formula

$$r(P) = \prod_{(i,j)\in P} r_{ij}, (5)$$

where r_{ij} is the reliability of the arc (i, j).

The problem of finding the most reliable path can be reduced to the problem of the shortest path, taking as the weight of the arc (i, j) and the value $c_{ij} = -\log r_{ij}$. Taking the logarithm from both parts of the equality in (5), we obtain

$$\log r(P) = \sum_{(i,j)\in P} \log r_{ij}.$$

Thus, for c_{ij} there is additivity, as in (4). It can be seen that the shortest path with the weight matrix $||c_{ij}||$ will be the most reliable path with the matrix $||r_{ij}||$; the reliability of this path is equal to the anti-logarithm of its length.

As such, by modifying the metrics (2) and (3) in this way, it is possible to find the controller locations that provide maximum reliability in their communication with other nodes (on average or for the worst case respectively).

Redundancy in SDNs

As mentioned above, the operability of an SDN controller (in the figures below: SDNC) and communication between the node and controller is necessary for the normal operation of the switch. Therefore, it is desirable that the reliability of the SDN switch, together with the controller and its means of communication, would not be lower than that of a traditional switch.

One of the most important methods for ensuring reliability is redundancy. We consider its application for an SDN in a typical example.

A. Description of a network example

As an example, we can take an SDN network as shown in Fig. 3. Its structure is chosen by analogy with the network considered in [34], being based on the national backbone network in Norway. The network under consideration consists of 11 nodes located in four major cities of the European part of Russia: Moscow (MSK), Saint Petersburg (SPB), Nizhny Novgorod (NN), and Rostov-on-Don (RoD). The nodes contain switches. Traffic from cities goes through access networks with connections to two, three or four nodes in the city.

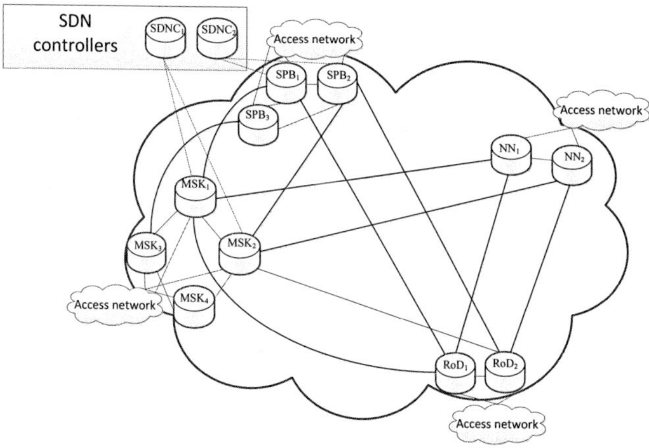

Fig. 3. Topology of the considered network

Three options for the location and connection of the controllers were considered:

- In the network there is one controller connected to one of the nodes in the city (in Moscow); only its communication with the node in another city has redundancy (Fig. 4).
- The same controller is connected to two nodes in the same city; each connection is duplicated separately with nodes in the other two cities, similar to the previous case (Fig. 5).
- In the network there are two controllers located in different cities (Moscow and St. Petersburg); each connection is duplicated (Fig. 6).

B. Calculations and results

The calculations were made based on the availability of the equipment in the nodes and links. The availability of SDN controllers and all switches in the nodes were taken as equal to 0.99999. The availabilities of links were calculated according to the well-known formula on the base of their mean operating time between failures (MTBF) and mean time to restoration (MTTR):

$$A = \frac{\text{MTBF}}{\text{MTBF} + \text{MTTR}}.$$

MTBF for a link with a length of L km, expressed in hours, can be calculated by the formula [18]

$$\text{MTBF}(L) = \frac{8760}{\nu \frac{L}{100}},$$

where 8760 = 24·365 is the number of hours per year and ν is the average number of failures per 100 km per year.

The following values for m and MTTR were taken from [18]: ν = 0.34; MTTR = 4.24 h.

If we consider the node in Rostov-on-Don, being the most distant from the cities where the controllers are located and its reliability will be lower.

The schemes shown in Figs. 4, 5, and 6 are presented as a combination of serial and parallel connections of the elements. Their availability was calculated according to the corresponding formulas in [31]. Further details can be found in [35]. The results of these calculations are given in Table I.

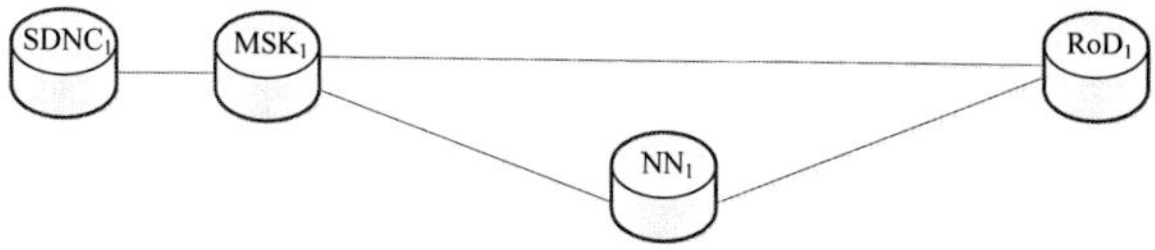

Fig. 4. The topology of the network fragment with a single SDN controller connected to a single node

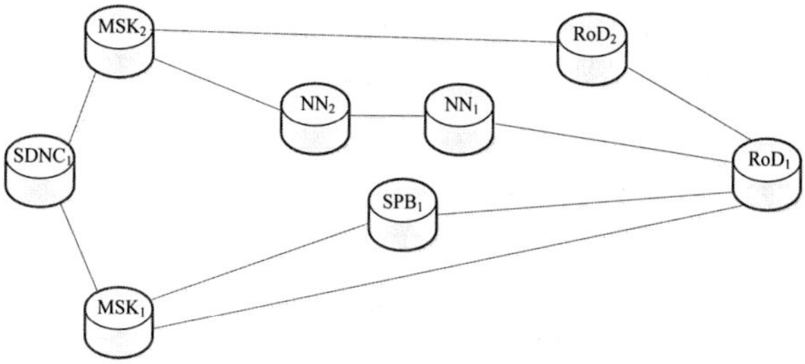

Fig. 5. The topology of the network fragment connecting one SDN controller to two nodes in the same city

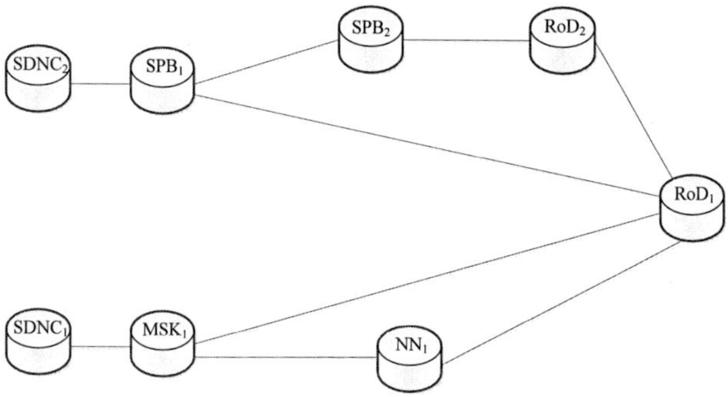

Fig. 6. The topology of the network fragment with two SDN controllers in different cities

Table 1. The availability of different options to connect SDN controllers

SDN controller(s) connection variant	Availability	Unavailability	Average downtime, minutes per year
1. One controller connected to one node	0.999965	$3.5 \cdot 10^{-5}$	18.40
2. One controller connected to two nodes in the same city	0.999979	$2.1 \cdot 10^{-5}$	11.04
3. Two controllers connected to two nodes in different cities	0.999989	$1.1 \cdot 10^{-5}$	5.78

Comparison of the results shows that the second variant has a downtime that is more than 1.5 times less than the first; while the third variant has a downtime that is more than 3 times less than the first. At the same time, only the network topology with two SDN controllers in different cities and the redundancy for paths from each node to controllers allows us to obtain a level of availability close to the required value of five nines.

The calculation results show the need for redundancy of controllers and their connections with all the nodes in the network.

The Problem of Calculating Network Reliability

The issue of reliability in the interaction of controllers with network equipment (southbound reliability), discussed above, is very important; but the main goal is to ensure reliability for end-to-end services from the perspective of end users. This requires an end-to-end reliability or availability assessment [13], [14]. There are a number of methods to do this, but SDNs pose a new challenge. Usually, in such calculations the states of network elements (nodes and links) are assumed to be statistically independent. In the case of an SDN, this assumption is not justified because the operability of multiple switches may depend on the same controllers. Besides this, the same links can be used for communication between terminals and for communication between switches and controllers.

Theoretically, such dependency can be taken into account by applying the method of power absorption, which has long been used in reliability theory [36], [37], [38], [39], [31]. In [36] and [31], this method was used to account for the existence of common elements belonging to different paths. In [37], it was used to account for node-failure probabilities. Its most general form is that of the general decomposition theorem, as proven in [38] and [39], and briefly presented in [31]. A number of previously known results, in particular those presented in [36] and [37], can be derived from this theorem as corollaries.

Using the method of power absorption, end-to-end network reliability taking into account the specificity of SDN can be calculated as described below.

The necessary notation will be introduced first. We assume that all the elements are numbered so that at the beginning of the list there will be nodes (switches), then links, then controllers. We denote the number of nodes, links, and controllers as n, l and m, respectively, $N = n + l + m$. Then the nodes take numbers from 1 to n, links take numbers from $n + 1$ to $n + l$, and controllers take numbers from $n + l + 1$ to N.

The first step is to express network reliability without taking into account the impact of controllers using any known method in the form

$$R = f(r_1, \dots, r_n, r_{n+1}, \dots, r_{n+l}), (6)$$

where the function f is a polynomial and r_i is the reliability of the ith network element.

In the second step for each node i $(i = 1, \dots, n)$ the probability of the existence of an operable controller that controls this node and its connection (as in the previous section) is expressed in the form

$$R_i = f_i(r_1, \dots, r_n, r_{n+1}, \dots, r_{n+l}, r_{n+l+1}, \dots, r_N), (7)$$

where f_i are also polynomials.

The final step involves substituting (7) for (6) and applying the operator of power absorption L to the resulting expression:

$$R = L[f(f_1(r_1, \dots, r_N), \dots, f_n(r_1, \dots, r_N), r_{n+1}, \dots, r_{n+l})]. (8)$$

The operator L means that in the open form of (8), which is a sum of the products, r_i should be substituted for any term of the form $r_i^{\alpha_i}$ if $\alpha_i > 1$. In [36] and [37], this operator is denoted by an asterisk (*).

More formally, an operator L can be defined as follows. Let f be an analytical function of k real variables having the form

$$f(x_1, \dots, x_k) = \sum_{\alpha_1,\dots,\alpha_k} a_{\alpha_1\dots\alpha_k} x_1^{\alpha_1} \dots x_k^{\alpha_k},$$

where $\alpha_i = 0,1, \dots$ and the number of summands is finite (in this case f is a polynomial) or infinite (if the series converges). Then

$$L[f(x_1, \dots, x_k)] = \sum_{\alpha_1,\dots,\alpha_k} a_{\alpha_1\dots\alpha_k} L(x_1^{\alpha_1}) \dots L(x_k^{\alpha_k}),$$

where $L(x_i^{\alpha_i}) = \begin{cases} 1, \text{if } \alpha_i = 0; \\ x_i, \ \text{if } \alpha_i \geq 1. \end{cases}$

Unfortunately, the described method is very cumbersome and time-consuming, so in practice it can only be directly applied to very small networks. Therefore, further research needs to be undertaken to develop more effective methods and suitable bounds.

Conclusion

The main findings of this paper are the following: the huge role of information and communication technologies in the life of modern society makes reliability a very important factor for communication networks; this fully applies to SDNs, forming the foundation of the digital economy and 5G network; achieving high reliability in an SDN requires redundancy, particularly for controllers and network connections; the cost of this redundancy must be taken into account when conducting a feasibility study for SDNs.

Future work could be devoted to the development of methods for the assessment of end-to-end network reliability. Typically, in such calculations, the states of network elements are assumed to be statistically independent. In the case of an SDN, this assumption is not justified because the operability of multiple switches may depend on the same controllers. Besides this, the same links can be used for communication between terminals and for communication between switches and controllers. This creates a serious problem that needs to be addressed further.

References

[1] P. Goranson, C. Black, T. Culver, Software Defined Networks: A Comprehensive Approach. Cambridge: Elsevier, 2017.
[2] ITU-T Recommendation Y.3300 (06/2014). Framework of Software-Defined Networking.
[3] V.A. Efimushkin, T.V. Ledovskikh, A.B. Ivanov, V.A. Shalaginov, "The role of SDN/NFV technologies in the digital economy infrastructure. Experience of testing and implementation", Elektrosviaz, 2018, No 3, pp. 27-36.
[4] Software-Defined Networking, https://en.wikipedia.org/wiki/Software-defined_networking#History.
[5] ITU-T Recommendation Y.3001 (05/2011). Future Networks: Objectives and Design Goals, https://www.itu.int/rec/T-REC-Y.3001-201105-I.
[6] ITU-T Recommendation Y.3101 (01/2018). Requirements of the IMT-2020 Network, https://www.itu.int/rec/T-REC-Y.3101.
[7] Recommendation ITU-T Y.3150 (01/2018). High-Level Technical Characteristics of Network Softwarization for IMT-2020, https://www.itu.int/rec/T-REC-Y.3150-201801-I.
[8] Recommendation ITU-T Y.3100 (09/2017). Terms and Definitions for IMT-2020 Network, https://www.itu.int/rec/T-REC-Y.3100-201709-I.
[9] S. Manning, "Future network. SDN consensus", https://www.opennetworking.org/news-and-events/blog/future-network-sdn-consensus/.
[10] V. Shalaginov, "Pilot tests of SDN solutions of communication operator data networks", in Proc. of the XI Internat. Conf. of "Information Society Technologies". Moscow: Media Publisher, 2017, pp. 425-426.
[11] Internet Technologies of the New Generation, http://en.arccn.ru/media/ons2017.
[12] IEC 60050-192:2015. International Electrotechnical Vocabulary – Part 192: Dependability.
[13] IEC 61907:2009. Communication Network Dependability Engineering.
[14] IEC 62673:2013: Methodology for Communication Network Dependability Assessment and Assurance.
[15] ITU-T Recommendation E.862 (06/92). Dependability Planning of Telecommunication Networks, https://www.itu.int/rec/T-REC-E.862-199206-I/en.
[16] T. Van Hardeveld, "International perspectives on reliability", https://reliabilityweb.com/articles/entry/international_perspectives_on_reliability.
[17] Theory of Communication Networks, Editor V.N. Rogisky. Moscow: Radio i Sviaz, 1981.

[18] V.P. Shuvalov, M.M. Egunov, E.A. Minina, Assurance of Reliability Measures for Telecommunication Systems and Networks. Moscow: Goryachay Liniya – Telecom. 2015.

[19] Cisco Systems White Paper. IP Telephony: The Five Nines Story, 2002, https://www.cisco.com/c/dam/global/it_it/solutions/pdf/ipcom/5nine_story_wp.pdf.

[20] W. Greene, B. Lancaster, "Carrier-grade: five nines, the myth and the reality", Pipeline, 2006, vol. 3, issue 11, http://www.pipelinepub.com/0407/pdf/Article%204_Carrier%20Grade_LTC.pdf.

[21] R. Stevenson, "How low-cost telecom killed five 9s in cloud computing", Wired, 2013, https://www.wired.com/insights/2013/03/how-low-cost-telecom-killed-five-9s-in-cloud-computing/

[22] 5 Nines Networks, http://www.5-nines.com/index.html.

[23] Achieving Five Nines of VNF Reliability in Telco-grade OpenStack Cloud. Panel discussion, April 27, 2016, https://www.openstack.org/assets/presentation-media/Five-Nine-Panel-Discussion-PA10.pdf.

[24] V.A. Netes, "Virtualization, cloud services and dependability", Vestnik Sviazy, 2016, No 8, pp. 7-9.

[25] V. Netes, "End-to-end availability of cloud services" in Proc. 22st Conf. of Open Innovations Association FRUCT, Jyvaskyla, Finland, May 2018, pp. 198-203.

[26] M. Ulema "Vulnerabilities and opportunities in SDN, NFV, and NGSON", in Proc. IEEE CQR 2014 Internat. Workshop. Emerging Technology Reliability Roundtable. Tucson, Arizona, USA, May 2014, pp. 24–27.

[27] N. Perrot, T. Reynaud. "Optimal placement of controllers in a resilient SDN Architecture", in Proc. 12th Internat. Conf. on the Design of Reliable Communication Networks (DRCN 2016). Paris, France, March 2016, pp. 145–151.

[28] B. Heller, R. Sherwood, N. McKeowin. "The controller placement problem", in Proc. ACM SIGCOM Workshop on Hot Topics in Software Defined Networking (HotSDN 2012), Aug. 2012, Helsinki, Finland, pp. 7-12.

[29] F.J. Ros, P.M. Ruiz, "Five nines of southbound reliability in Software-Defined Networks", in Proc. ACM SIGCOMM Workshop on Hot Topics in Software Defined Networking (HotSDN 2014), Aug. 2014, Chicago, IL, USA, pp. 31-36.

[30] Jiang Liu, Juan Liu, R. Xie, "Reliability-based controller placement algorithm in Software Defined Networking", Computer Science and Information Systems, 2016, vol. 13, No 2, pp. 547-560.

[31] Handbook of Reliability Engineering, Editor I.A. Ushakov. New York: John Wiley & Sons, 1994.

[32] H. Frank, I.T. Frisch, Communication, Transmission, and Transportation Networks. Reading: Addison Wesley, 1971.

[33] N. Christofides, Graph Theory: an Algorithmic Approach. London, New York: Academic Press, 1975.

[34] G. Nencioni, B.E. Helvik, A.J. Gonzalezy, P.E. Heegaard, A. Kamisinski, “Impact of SDN controllers deployment on network availability”, Department of Telematics, Norwegian University of Science and Technology, Trondheim, Norway, 2017.

[35] V.A. Netes, M.S. Kusakina, “Reliability of communication between controllers and switches in SDN”, Vestnik Sviazy, 2018, No 9, pp. 10-13.

[36] Y.H. Kim, K.E. Case, P.M. Ghare, “A method for computing complex system reliability”, IEEE Trans. on Reliability, 1972, vol. R-21, No 4, pp. 215-219.

[37] K.K. Aggarwal, J.S. Gupta, K.B. Misra, “A simple method for reliability evaluation of a communication network”, IEEE Trans. on Communications, 1975, vol. COM-23, No 5, pp. 563-566.

[38] V.A. Netes, “Expected value of the effectiveness of discrete systems”, Automatic Control and Computer Sciences, 1980, vol. 14, No 3, pp. 68-71.

[39] V.A. Netes, “Use of a linear representation of the efficiency function for calculating it”, Engineering Cybernetics, 1982, vol. 20, No 2, pp. 77-84.

CHAPTER 1.3

USER TRAFFIC HANDLING FOR MISSION CRITICAL COMMUNICATIONS AT THE NETWORK EDGE

EVELINA PENCHEVA, IVAYLO ATANASOV, DENITSA VELKOVA, VENTSISLAV TRIFONOV

Abstract. Multi-access Edge Computing (MEC) is a promising technology that may satisfy the diverse service requirements of 5G services. It offers cloud capabilities for storage and computing in proximity to end users and thus enables a more timely reaction to dynamic network conditions. MEC can embrace the challenges posed by mission critical applications, providing required levels of availability, reliability, latency, and security. In this chapter, we propose a new mobile edge service, which provides functionality for user traffic handling. Using the service interface, a mobile edge application may shape, redirect, or block user traffic according to information on radio conditions, user location, and requested data speeds, etc. This new mobile edge service is described for its typical use cases, data model, interface definition, and state models, which are formally verified. In addition, different MEC deployment scenarios are studied and its latency is evaluated.

Introduction

Fifth generation (5G) networks are expected to see improved mobile network performance, including quality of service (QoS). The diverse QoS requirements of different services are a challenge for network operators, as telcos have to take into consideration the limited radio frequency spectrum and backhaul capacity [1], [2]. The concept of a flexible radio access network with cloud intelligence is a viable solution and offers a compromise between increasing capacity and total infrastructure costs [3].

Multi-access Edge Computing (MEC) is a technology that provides cloud intelligence at the network's edge, as well as distributed computing and storage capabilities. Closer vicinity to end users reduces the latency to a few milliseconds and makes MEC suitable for mission critical communications. The presence of cloud intelligence at the network's edge is seen as an absolute necessity to enable mission critical communications, requiring low end-to-end latency [4].

MEC is based on the virtualization of network functions. These network functions include the control and management of QoS, service policies, and traffic prioritization [5], [6]. MEC provides flexibility to achieve QoS isolation for customized 5G services through network slicing [7], [8].

According to MEC technical requirements, the mobile edge platform, or the proposed mobile edge application, should offer the possibility of an authorized application to statically and/or dynamically register its bandwidth and/or priority requirements, as well as allocate bandwidth and priority to each session or application [9].

A bandwidth management service (BWMS), as defined by the European Telecommunications Standard Institute (ETSI), provides a means to efficiently and rapidly address the requirements of bandwidth and/or priority of various mobile edge applications or sessions of the same application [10]. BWMS application programming interfaces (APIs) allow an authorized mobile edge application to allocate bandwidth and/or assign priority to any session or any application. A BWMS can merge all bandwidth management requests and optimize usage.

In this chapter, we propose a new mobile edge service that provides a mechanism for dynamic user traffic handling and thus for providing packet filters for selected traffic. The novelty of this research is that it offers the possibility for a mobile edge platform or authorized mobile edge application to inspect, shape, redirect, or block user traffic according to policy rules. The traffic handling rules may depend on network congestion, user location, and QoS requirements; or they may be defined according to the mobile edge application.

We also study two possible MEC deployment scenarios, as presented in [11]. One of these deployment scenarios co-locates MEC and a cloud radio access network (C-RAN), while the other uses MEC in distributed core network functionality. Signaling procedures in a radio access network related to MEC functionality for the provision of real-time radio network information and the installation of rules for user traffic handling are considered for both deployment scenarios. Performance analysis is given in terms of latency.

Research Motivation

Information on current radio conditions is shared through the MEC platform via a radio network information service (RNIS) [12]. A RNIS is a service that provides radio-related information for mobile edge applications and mobile edge platforms. The granularity of radio network information can be set according to various parameters, such as cell, user equipment (UE), and service quality class; or they can be requested over a period of time. Typical information that can be provided, includes:

- up-to-date information on radio network conditions;
- measurements related to the user plane;
- information and changes related to UEs, served by radio node(s) connected to the MEC server, including radio resources allocated to UEs.

An authorized mobile edge application may receive cell level radio access bearer (RAB) information and may subscribe to receive notifications about RAB establishment, modification, and release. Upon receiving a request for RAB management, the mobile edge application is notified.

According to the technical requirements of the MEC, essential traffic routing properties include the following [9]:

- the MEC platform must provide functionality for routing selected user traffic in the uplink and/or downlink directions to an authorized mobile edge application;
- the MEC platform must provide functionality for inspecting, modifying, and shaping user traffic in uplink and/or downlink directions;
- MEC management must allow the configuration of traffic rules. Traffic rules can allow the establishment of packet filters based on a network address and/or IP protocol based on a tunnel endpoint identifier and/or a subscriber profile identifier, and/or a quality class identifier.

The purpose of the UE identity service is to allow for UE-specific traffic rules in the mobile edge system [13]. The UE identity API provides functionality for a mobile edge application to register a tag (representing a UE) or a list of tags. A UE identity tag's registration triggers the mobile edge platform to activate the corresponding traffic rule(s) linked to the tag. Later, if the application does not wish to use the traffic rule for that user, it may de-register the UE identity tag by invoking the de-registration procedure. The current standard version does not provide a mechanism for the realization of tag mapping onto a specific UE in the mobile network operator's system. Also, it does not describe information for the related application policy, e.g. traffic rule patterns.

In this paper, a new mobile edge service, a user traffic handling service (UTHS), is presented based on MEC routing requirements. The service enables mobile edge applications to manage user traffic. Depending on the policy provided to the mobile edge application's logic, the following actions can be performed on user traffic: continue with existing data; traffic blocking or traffic shaping in uplink and/or downlink directions; as well as user traffic redirection to another destination. Special user traffic handling is a very important feature for mission critical communications, which have specific QoS requirements.

Information Service Description

The proposed UTHS allows authorized mobile edge applications to:

- inspect selected uplink and/or downlink user traffic;
- shape selected uplink and/or downlink user plane traffic;
- select routes for uplink and/or downlink user traffic from the network to the authorized mobile edge application;
- block selected uplink and/or downlink user traffic.

Decisions on handling user traffic may depend on current user location, radio network conditions, or service (e.g., for mission critical services, the UTHS may grant premium bit rates).

A use case for UTHS may be that the user has a subscription for a fixed mobile service at home where he/she receives preferential QoS values. When the user leaves their home area, the user traffic is shaped. Actual information about user location may be provided by the mobile edge location service [14]. In order to receive location information, the mobile edge application subscribes to a location service to receive notifications on the user entering or leaving a specified area.

Another use case for UTHS is based on actual radio network information. An analytical mobile edge application subscribes to an RNIS for the setup, modification, and release of RABs dedicated to a specific UE, as well as for handover related events (cell changes). Having information about current radio access network conditions, the application may initiate actions related to the user traffic e.g. shaping, redirecting, or blocking. Reasons for these actions may include congestion of the radio access network, or exhaustion of available user credit, among others.

The proposed UTHS may be used in conjunction with a mobile edge service, which provides functionality for usage monitoring control and application detection and control. Both control types are part of the 5G Policy and Charging Control Framework. Usage monitoring control enables the monitoring and reporting of the accumulated usage of network resources on a per session basis and per user basis. In [15], APIs that enable the enforcement of dynamic policy decisions based on total network usage in real-time at the network edge are defined. The application detection and control (ADC) feature enables detection of specified application traffic, reporting on the start and stop of application traffic, and applying specific enforcement actions for detected application traffic. Distributing ADC functionality at the network edge allows for a timely response to events concerning data traffic.

The following section describes the data model related to the proposed service.

Data Model

Data structures related to the user traffic handling actions include the following: **ShapingInfo**, **GatingInfo** and **RedirectionInfo.**

The **ShapingInfo** data structure describes bandwidth limitation information. This is a structure with the following items:

- **mBitRateDl** represents the maximum downlink bit rate and is of integer type;
- **mBitRateUl** represents the maximum uplink bit rate and is of integer type;

- **gBitRateDl** represents the guaranteed downlink bit rate and is of integer type;
- **gBitRateUl** represents the guaranteed uplink bit rate and is of integer type;
- **5Gguti** is an unambiguous identification of the UE. It does not reveal the UE or the user's permanent identity in the 5G system. It also allows the identification of the access and mobility management function (AMF) and network. This is a structure of **mcc, mnc, amfID** and **5Gtmsi**;
- **mcc** is the mobile country code;
- **mnc** is the mobile network code;
- **amfID** is the identification of the AMF serving the UE;
- **5Gtmsi** is the temporary mobile subscriber identity;
- **shapingDuration** represents the duration of the rate shaping in seconds.

The **BlockingInfo** data structure describes information about user traffic blocking. This is a structure of **direction, 5Gguti** (as described above) and **blockingDuration**, where:

• **direction** indicates the user traffic direction that should be blocked and is of enumerated type: 0 = downlink, 1 = uplink, 2 = downlink and uplink;

• **blockingDuration** represents the duration of user traffic blocking.

The **RedirectionInfo** data type describes information about user traffic redirection. This is a structure of **redirectServerAddress, 5Gguti** (as described above) and **redirectDuration**, where:

• **redirectServerAddress** represents the address of the server to which the user traffic must be redirected;

• **redirectDuration** represents the duration of the redirection of application traffic in seconds.

The motivation behind the use of the data model is that the different applications store and access the same data structures, enabling interoperability.

API Definition

The resource structure supported by the UTHS is shown in Fig. X-1.

Each resource has a unique uniform resource identification (URI). All UTHS resources have the following root:

{apiRoot}/uths/{apiVersion}/

Table I presents the resources and supported methods.

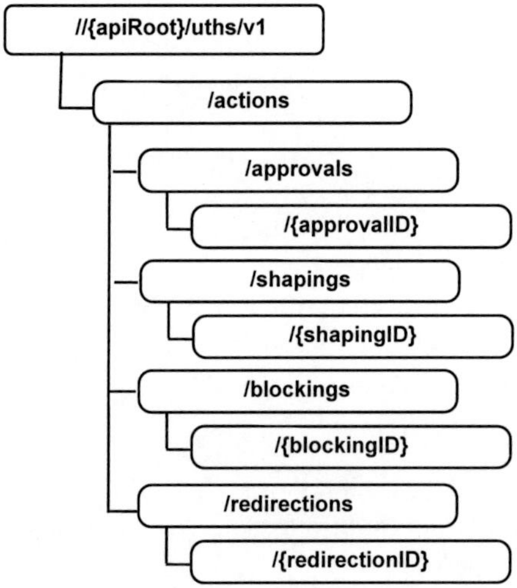

Fig.1 Structure of resources supported by UTHS

Following the RESTful architectural style, all resources are manipulated using four simple operations: CREATE (HTTP POST request); READ (HTTP GET request); UPDATE (HTTP PUT request); and DELETE (HTTP DELETE request).

In the next section, we consider some issues regarding implementation of the proposed mobile edge service.

State Models

Implementation of the mobile edge UTHS and respective mobile edge applications in the network requires the development of models representing RAB states. The models representing the state of a RAB for a given UE are supported by the MEC platform, the serving gNB (5G base station), and a mobile edge application needing to be synchronized.

Table 1. Resources and supported HTTP methods

Resource name	Resource URI	HTTP method	Description
All actions on user traffic	/actions	GET	Retrieves the list of all enforcement actions.
All approvals of RAB setup or modification	/actions/ approvals	GET POST	Retrieves a list with all RAB setup or modification requests approved by the application. Creates a new approval of RAB setup or modification.
Existing action related to approval of RAB setup or modification	/actions/ approvals/ approvalID	GET PUT DELETE	Retrieves information about current approval of RAB setup or modification. Modifies existing approval of RAB setup or modification. Deletes existing approval of RAB setup or modification.
All actions related to user traffic shaping	/actions/ shapings	GET POST	Retrieves a list with all actions related to user traffic shaping. Creates a new action related to user traffic shaping.
Existing action related to traffic shaping	/actions/ shapings/ shapingID	GET PUT DELETE	Retrieves information about existing user traffic shaping. Modifies existing user traffic shaping. Deletes existing user traffic shaping.
All actions related to user traffic redirection	/actions/ redirections	GET POST	Retrieves a list with all actions related to user traffic redirections. Creates a new action related to user traffic redirections.
Existing action related to traffic redirection	/actions/ redirections/ redirectionID	GET PUT DELETE	Retrieves information about existing user redirection. Modifies existing user traffic redirection. Deletes existing user traffic redirection.
All actions related to user traffic blocking	/actions/ blockings	GET POST	Retrieves a list with all actions related to user traffic blocking. Creates a new action related to user traffic blocking.

Existing action related to traffic blocking	/actions/ blockings/ blockingID	GET PUT DELETE	Retrieves information about existing user traffic blocking. Modifies existing user traffic blocking. Deletes existing user traffic blocking.

Let us consider an authorized mobile edge application that subscribes to events related to a RAB setup, modification, and release. When an RAB management event occurs, the application is notified and may initiate user traffic shaping, redirection, or blocking. The simplified model, representing the mobile edge application's view on the RAB state is presented in Fig. X-2.

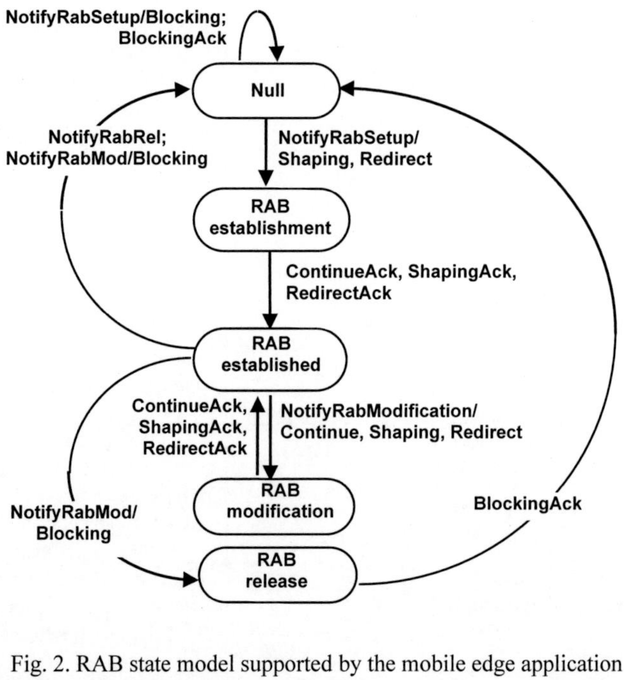

Fig. 2. RAB state model supported by the mobile edge application

The RAB management services are related to one UE and use a UE-associated signaling connection that is maintained for the UE in question. They include the Packet Data Unit (PDU) Session Resource Setup; the PDU Session Release; and the PDU Session Modify procedure [16]. The purpose of the PDU Session Resource Setup procedure is to assign resources

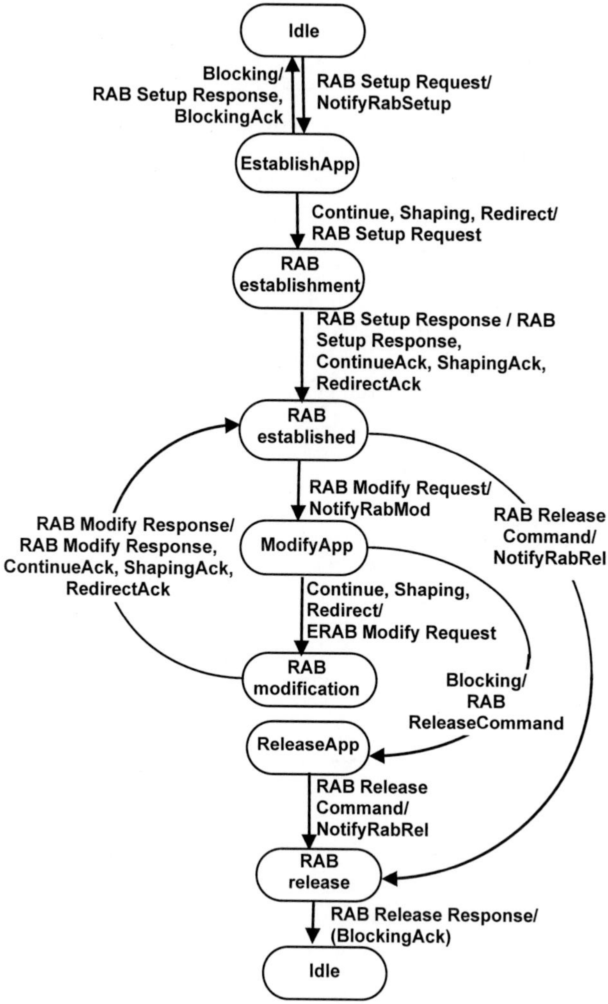

Fig. 3. RAB state model supported by MEC platform

in a RAN for one or more PDU session resources and corresponding QoS flows; as well as to set up corresponding data radio bearers for a given UE. The purpose of the PDU Session Resource Release procedure is to enable the release of already established PDU session resources for a given UE. The purpose of the PDU Session Resource Modify procedure is to change the configuration of an already established PDU session(s) for a given UE.

This also enables the setup, modification, and release of the QoS flow for an already established PDU session(s). If the application has an active subscription, in case of a PDU Session Resource Setup/Modify request, the MEC platform notifies the application about the event and expects application instructions on how to proceed. The mobile edge application may in turn approve PDU session resource setup/modification, or may initiate user traffic handling actions. The simplified model representing the RAB state model supported by the MEC platform is presented in Fig. 3.

A simplified model representing the view of the serving gNB on the RAB state is illustrated in Fig. 4. The RAB setup, modification, and release procedures result in the reconfiguration or release of the radio resource control (RRC) connection [17].

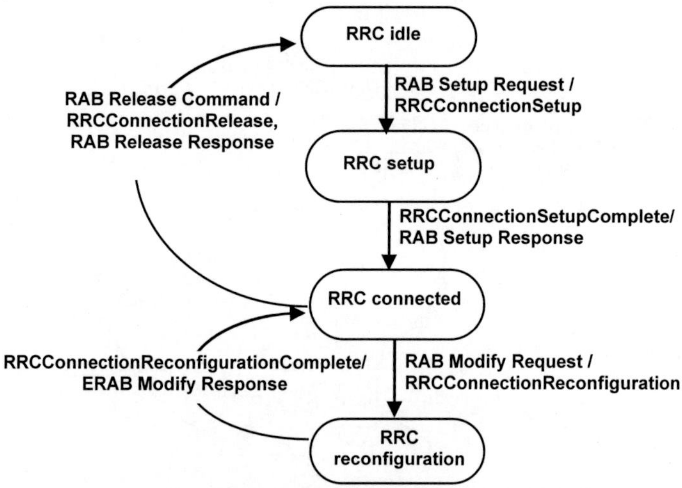

Fig. 4. RAB state model supported by gNB

The three models are simplified, only representing successful execution of the relevant procedures in the network, and they do not take into account abnormal conditions associated with unsuccessful completion.

In order to provide a more rigorous proof that the three models expose equivalent behavior, we formalize the descriptions of the modes. Descriptions of the RAB state are formalized using the notion of a labeled transition system (LTS).

A labeled transition system is represented as a quadruple: a set of states; a set of actions; a set of transitions; and a set of initial states.

T_{App}= (S_{App}, Act_{App}, $\rightarrow_{App}$, s_0^{App}) denotes an LTS, representing the RAB state model supported by a mobile edge application, where:

- S_{App} = {Null [s_1^A], RABestablishment [s_2^A], RABestablished [s_3^A], RABmodification [s_4^A], RABrelease [s_5^A]};

- Act_{App} = {NotifyRabSetup [t_1^A], ContinueAck [t_2^A], ShapingAck [t_3^A], RedirectAck[t_4^A], NotifyRabMod [t_5^A], BlockingAck [t_6^A], NotifyRabRel [t_7^A]};

- $\rightarrow_{App}$ = {(s_1^A t_1^A s_1^A), (s_1^A t_6^A s_1^A), (s_1^A t_1^A s_2^A), (s_2^A t_2^A s_3^A),(s_2^A t_3^A s_3^A), (s_2^A t_4^A s_3^A), (s_3^A t_5^A s_4^A), (s_4^A t_2^A s_3^A), (s_4^A t_3^A s_3^A), (s_4^A t_4^A s_3^A), (s_3^A t_5^A s_5^A), (s_5^A t_6^A s_1^A), (s_3^A t_7^A s_1^A), (s_3^A t_5^A s_1^A)};

s_0^{App} = {s_1^A}.

Short notations for states and actions are given in brackets.

T_P = (S_P, Act_P, $\rightarrow_P$, s_0^P) denotes an LTS, representing the RAB state model supported by the MEC platform, where:

- S_P = {Idle [s_1^P], EstablishApp [s_2^P], RABestablishment [s_3^P], RabEstablished [s_4^P], ModifyApp [s_5^P], RABmodification [s_6^P], ReleaseApp [s_7^P], RABrelease [s_8^P]};

- Act_P = {RABSetupRequest [t_1^P], Blocking [t_2^P], Continue [t_3^P], Shaping [t_4^P], Redirect [t_5^P], RABSetupResponse [t_6^P],

RABModifyRequest [t_7^P], RABModifyResponse [t_8^P],

RABReleaseCommand [t_9^P], RABReleaseResponse [t_{10}^P]};

- $\rightarrow_P$ = {(s_1^P t_1^P s_2^P), (s_2^P t_2^P s_1^P), (s_2^P t_3^P s_3^P), (s_2^P t_4^P s_3^P),
(s_2^P t_5^P s_3^P), (s_3^P t_6^P s_4^P), (s_4^P t_7^P s_5^P), (s_5^P t_3^P s_6^P),
(s_5^P t_4^P s_6^P), (s_5^P t_5^P s_6^P), (s_6^P t_8^P s_4^P), (s_4^P t_9^P s_8^P),
(s_5^P t_2^P s_7^P), (s_7^P t_9^P s_8^P), (s_8^P t_{10}^P s_1^P)};

s_0^P = {s_1^P}.

$T_{gNB} = (S_{gNB}, Act_{gNB} \rightarrow_{gNB}, s_0^{gNB})$ denotes an LTS, representing the RAB state model supported by gNB, where:

- S_{gNB} = {RRCidle [s_1^N], RRCsetup [s_2^N], RRCconnected [s_3^N],
RRCreconfiguration [s_4^N]};

- Act_{gNB} = {RABSetupRequest [t_1^N], RRCConnectionSetupComplete
[t_2^N], RABModifyRequest [t_3^N],
RRCConnectionReconfigurationComplete [t_4^N],
RABReleaseCommand [t_5^N]};

- $\rightarrow_{gNB}$ = {(s_1^N t_1^N s_2^N),(s_2^N t_2^N s_3^N), (s_3^N t_3^N s_4^N), (s_4^N t_4^N s_3^N),
(s_3^N t_5^N s_1^N)};

- s_0^{gNB} = {s_1^N}.

Having defined the descriptions of the formal models, it can be proved that the three models expose equivalent behavior.

Intuitively, in terms of observed behavior, two LTSs are equivalent if one LTS displays a final result and the other LTS displays the same result. The idea of equivalence is formalized by the concept of bisimilarity [18]. In practice, strong bisimilarity leads to strong conditions for equivalence, which are not always necessary. Weak bisimilarity causes internal transitions to be ignored.

Proposition: T_{App}, T_{P}, and T_{gNB} are weakly bisimilar.

Proof: As to the definition of weak bisimilarity, it is necessary to identify a relation between the states of the three LTSs, as for any transition from a state in one LTS there are respective transitions from states in the other LTSs.

U_{AppPgNB} denotes a relation between the states of T_{App}, T_{P}, and T_{gNB}, where $U_{\mathrm{AppPgNB}} = \{(s_1^A, s_1^P, s_1^N), (s_3^A, s_4^P, s_3^N)\}$. Then, for each of the following events the respective transitions for the states in U_{AppPgNB} are identified:

1. A mobile edge application receives a notification about the setup of an RAB, dedicated to user traffic, and it approves the RAB setup for: $(s_1^A\ t_1^A\ s_2^A), (s_2^A\ t_2^A\ s_3^A) \exists \{(s_1^P\ t_1^P\ s_2^P), (s_2^P\ t_3^P\ s_3^P), (s_3^P\ t_6^P\ s_4^P)\} \sqcup \{(s_1^N\ t_1^N\ s_2^N), (s_2^N\ t_2^N\ s_3^N)\}$.

2. A mobile edge application receives a notification about the setup of an RAB dedicated to user traffic. It initiates traffic shaping for: $(s_1^A\ t_1^A\ s_2^A), (s_2^A\ t_3^A\ s_3^A) \exists \{(s_1^P\ t_1^P\ s_2^P), (s_2^P\ t_4^P\ s_3^P), (s_3^P\ t_6^P\ s_4^P)\} \sqcup \{(s_1^N\ t_1^N\ s_2^N), (s_2^N\ t_2^N\ s_3^N)\}$.

3. A mobile edge application receives a notification about the setup of an RAB dedicated to user traffic. It initiates traffic redirection for: $(s_1^A\ t_1^A\ s_2^A), (s_2^A\ t_4^A\ s_3^A) \exists \{ (s_1^P\ t_1^P\ s_2^P), (s_2^P\ t_5^P\ s_3^P), (s_3^P\ t_6^P\ s_4^P)\} \sqcup \{(s_1^N\ t_1^N\ s_2^N), (s_2^N\ t_2^N\ s_3^N)\}$.

4. A mobile edge application receives a notification about the setup of an RAB dedicated to user traffic. It initiates traffic blocking for: $(s_1^A\ t_1^A\ s_1^A)$, $(s_1^A\ t_6^A\ s_1^A) \exists (s_1^P\ t_1^P\ s_2^P)$, $(s_2^P\ t_2^P\ s_1^P)$.

5. A mobile edge application receives a notification about the modification of an RAB dedicated to user traffic and the application approves the RAB modification for: $(s_3^A\ t_5^A\ s_4^A)$, $(s_4^A\ t_2^A\ s_3^A) \exists$ $\{(s_4^P\ t_7^P\ s_5^P)$, $(s_5^P\ t_3^P\ s_6^P)$, $(s_6^P\ t_8^P\ s_4^P)\} \sqcup \{(s_3^N\ t_3^N\ s_4^N)$, $(s_4^N\ t_4^N\ s_3^N)\}$.

6. A mobile edge application receives a notification about the modification of an RAB dedicated for user traffic and it initiates traffic shaping for: $(s_3^A\ t_5^A\ s_4^A)$, $(s_4^A\ t_3^A\ s_3^A) \exists \{(s_4^P\ t_7^P\ s_5^P)$, $(s_5^P\ t_4^P\ s_6^P)$, $(s_6^P\ t_8^P\ s_4^P)\} \sqcup \{(s_3^N\ t_3^N\ s_4^N)$, $(s_4^N\ t_4^N\ s_3^N)\}$.

7. A mobile edge application receives a notification about the modification of an RAB dedicated to user traffic. It initiates traffic redirection for: $(s_3^A\ t_5^A\ s_4^A)$, $(s_4^A\ t_4^A\ s_3^A) \exists \{(s_4^P\ t_7^P\ s_5^P)$, $(s_5^P\ t_5^P\ s_6^P)$, $(s_6^P\ t_8^P\ s_4^P)\} \sqcup \{(s_3^N\ t_3^N\ s_4^N)$, $(s_4^N\ t_4^N\ s_3^N)\}$.

8. A mobile edge application receives a notification about the modification of an RAB dedicated to user traffic. It initiates traffic blocking for: $(s_3^A\ t_5^A\ s_1^A)$, $(s_1^A\ t_6^A\ s_1^A) \exists \{(s_4^P\ t_7^P\ s_5^P)$, $(s_5^P\ t_2^P\ s_7^P)$, $(s_7^P\ t_9^P\ s_8^P)$, $(s_8^P\ t_{10}^P\ s_1^P)\} \sqcup (s_3^N\ t_5^N\ s_1^N)\}$.

9. A mobile edge application receives a notification about the release of an RAB dedicated to user traffic for: $(s_3^A\ t_7^A\ s_1^A) \exists \{(s_4^P\ t_9^P\ s_8^P)$, $(s_8^P\ t_{10}^P\ s_1^P)\} \sqcup (s_3^N\ t_5^N\ s_1^N)$.

Therefore, T_{App}, T_P, and T_{gNB} are weakly bisimilar, i.e. they display equivalent behavior.

In the following sections, we study different MEC deployment scenarios in the 5G system.

Mobile Edge Host Placements in 5G System Architecture

The mobile edge host is an entity containing a mobile edge platform and a virtualization infrastructure, which provides computing, storage, and network resources, for the purpose of running mobile edge applications. The placement of the mobile edge host in the 5G system architecture is important in terms of latency.

The greatest benefit of MEC technology can be found when the mobile edge host is located as close as possible to the network's edge, e.g. in the 5G next generation radio access network (NG-RAN). The best placement of the mobile edge host running services as an RNIS, which may be used to optimize the services offered to the UE, is at the gNB [11].

Another scenario, which is appropriate for enterprise networks and mission critical communications, involves distributed core network functionality. In this case, core network functions or a part of them dedicated to the needs of local private networks, are distributed at the network's edge, and the mobile edge host is placed in the vicinity of this private network.

In 5G NG-RAN, the monolithic architecture of gNB is split into one centralized unit (gNB-CU), which supports upper layer protocols, and one or more distributed units (gNB-DUs), which support lower layer protocols. The F1 interface between gNB-CU and gNB-DU supports signaling exchange and data transmission, separating the radio network layer and the transport network layer, and enabling the exchange of UE-associated and non-UE-associated signaling [19]. The split between gNB-CU and gNB-DU helps to virtualize network functions increasing flexibility and leading to cost reductions. In the case of gNB-CU and MEC host co-location, mobile edge services and applications can share the same virtualized infrastructure as virtualized gNB-CU functions. This deployment scenario is shown in Fig. 5.

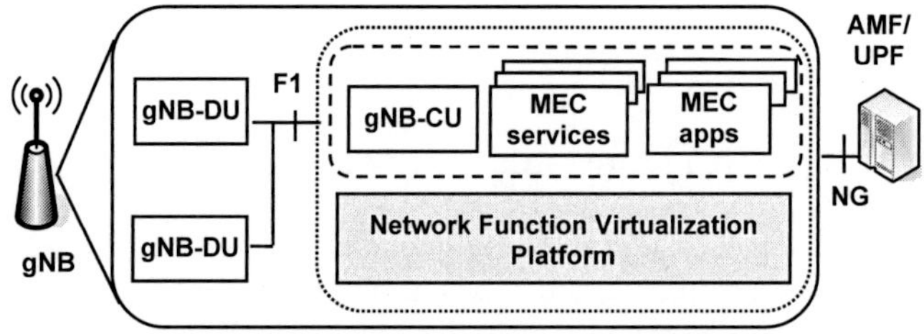

Fig. 5. MEC and gNB-CU co-location

The NG interface includes the gNBs, the AMF, and the user plane function (UPF).

The 5G core network is designed to meet the requirements of different use cases and features service-based architecture. The core network functions are defined as services using standard APIs. They can run as virtualized network functions in order to reduce costs. Mobile edge services and applications can be co-located with virtualized core network functions on the same virtualized platform, improving scalability and resulting in better utilization of network resources. The deployment scenario with distributed core network functions is shown in Fig. 6.

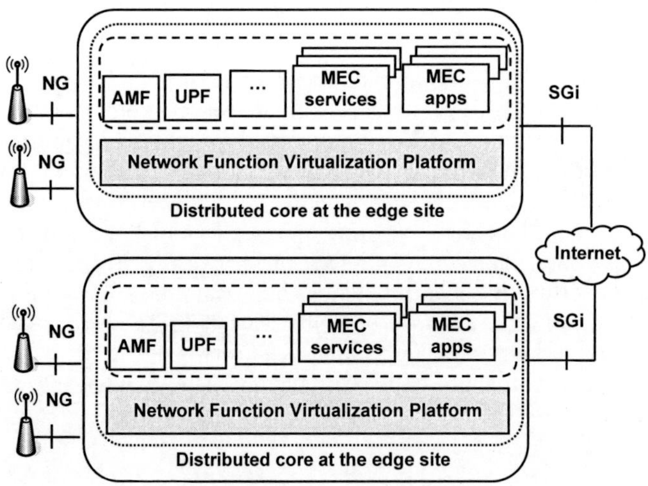

Fig. 6. MEC and co-location of distributed core network functions

In the following section, we consider signaling scenarios for both mobile edge host deployment types in providing mobile edge RNIS and UTHS.

Signaling for Mobile Edge Services

In order to evaluate latency, we modeled the signaling of mobile edge services for both mobile edge host deployment scenarios. The signaling protocol between gNB-DU and gNB-CU is F1AP (F1 application protocol) [20]. The NG application protocol (NGAP) provides the signaling service between the gNB-CU and the AMF [16]. The radio resource control (RRC) protocol is used for the on-the-air interface [17].

Fig. 7 shows the message flow in the provisioning of radio network information for the RAB setup in the case of a MEC and gNB-CU co-location. The interaction between the gNB-CU and MEC platforms, providing RNIS, is internal. Short notations of messages are given at message ends in brackets.

A precondition is that the mobile edge application has an active subscription with RNIS for notification of RAB management events. The sequence of steps is as follows:

1) The AMF requests setup of the PDU session resource.

2) The gNB-CU transfers the request for the PDU Session Resource Setup to the gNB-DU.

3-4) The RRC connection is reconfigured in order to set up the requested RABs.

5) The gNB-DU transfers the response to the gNB-CU.

6-7) The RNIS sends a notification to the mobile edge application for RAB establishment.

8) A PDU Session Resource Setup response is sent to the AMF.

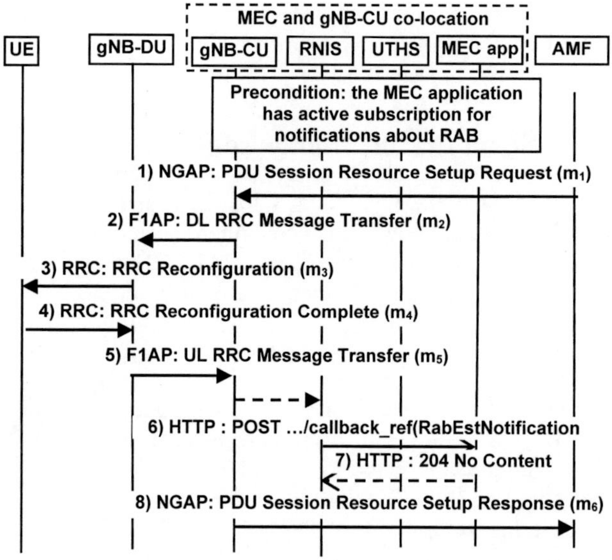

Fig. 7. Flow for provisioning information about an RAB setup: MEC and gNB-CU co-location deployment

Fig. 8 shows the flow for the PDU session resource setup in the case of MEC and distributed core functions at the edge site.

The difference between MEC and gNB-CU collocation deployment is that the information about RAB establishment needs to be reported to the AMF first, which in turn provides information through internal communication to the respective mobile edge services.

A similar flow pattern is applied for notifications about RAB modification and release.

Fig. 9 shows the flow for traffic shaping by an authorized mobile edge application in the case of MEC and gNB-CU co-location.

Actions related to user traffic redirection and release follow the same message exchange pattern.

The next section provides an assessment of latency introduced by the control plane for the scenarios described above.

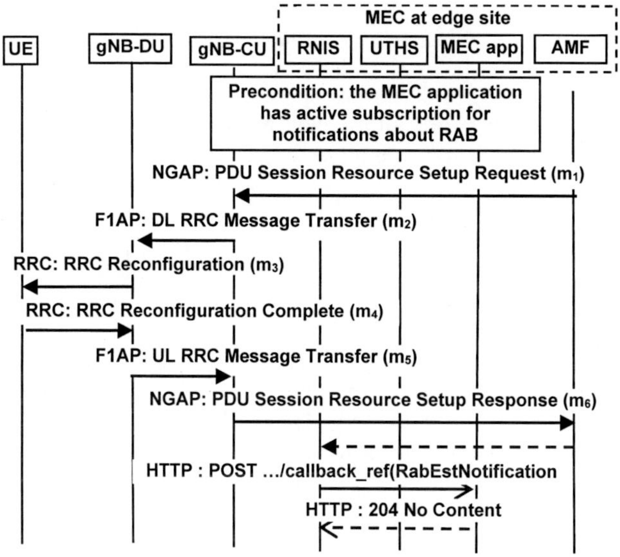

Fig. 8. Flow for provisioning information about RAB setup: MEC with distributed core functions

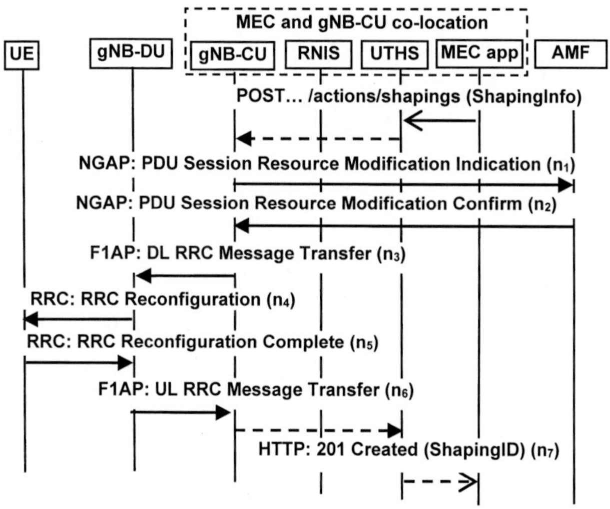

Fig. 9. Flow for user traffic shaping by mobile edge application: MEC and gNB-CU co-location deployment

Latency Assessment

The total latency introduced by the control plane protocols in provisioning radio network information may be calculated as the sum of times required for the transmission of protocol messages over interfaces and the sum of processing times in each node. The maximum requirement for control plane latency is 20 ms. Proponents are encouraged to consider lower control plane latency, e.g. 10 ms. As in [21], [22], the NG-RAN node/UE processing time is about 0.25-0.3 ms and the time required for message transfer is 0.1 ms.

α denotes the time required for message transfer and β denotes the UE/NG-RAN node processing time. Latency in the provisioning of radio network information to the mobile edge application may be calculated as the sum of all times required for message transfer over NGAP, F1, and air interfaces, and the sum of processing times in the NG-RAN node and UE. The values of $\alpha = 0.1$ and $\beta = 0.3$ are used for coarse-grained latency evaluation in the provisioning of radio network information. The time cost for MEC execution is given as

$$\beta(\text{MEC}) = \text{D.X/f}$$

where *D* is information on the task input-data size *D* (in bits); *X* is the computation workload/intensity (in CPU cycles per bit); and *f* is the CPU frequency of the MEC server. The MEC parameters used are as follows: input data size *D* = 1,000 bits; computational intensity *X* 1,000 cycles/bit; and the CPU frequency of the mobile edge host *f* 1.1 GHz [23].

The time budget T(rni) for the case of the MEC and gNB-CU co-location scenario, shown in Fig. 7, is as follows:

$$\text{T(rni)} = \alpha\ (\text{NGAP}(m_1)) + \beta(\text{gNB-CU}) + \alpha(\text{F1AP}(m_2)) + + \beta(\text{gNB-DU}) + \alpha\ (\text{RRC}(m_3)) + \beta(\text{UE}) + \alpha(\text{RRC}(m_4) + \beta(\text{gNB-DU} + \alpha(\text{F1AP}(m_5)) + \beta(\text{gNB-CU}) + 2\beta(\text{MEC}) + \alpha(\text{NGAP}(m_6)) + 2\beta(\text{AMF}) = 6\alpha + 6\beta + 2\beta(\text{MEC}) = 2.4\text{ ms} + + 2\beta(\text{MEC}) = 2.4\text{ ms} + 1.82\text{ ms} = 4.22\text{ ms.}$$

The time budget for the case of MEC with distributed core functions is the same. The difference between the two deployment scenarios is that the latency in provisioning radio network information to the MEC in the case of MEC and gNB co-location is less than that for MEC with distributed core functions by 0.4 ms, as the information is provided before sending the NGAP: the PDU Session Resource Setup response to the core network.

The latency reduction in the case of MEC and gNB-CU co-location is a result of the internal interface between the gNB-CU and MEC.

The time budget T(uth) for performing mobile edge application initiated actions on user traffic, shown in Fig. 9, is as follows:

$$T(uth) = \beta(MEC) + \alpha(NGAP(n_1)) + \beta(AMF) + \alpha(NGAP(n_2)) + \beta(gNB\text{-}CU) + \alpha(F1AP(n_3)) + \beta(gNB\text{-}DU) + \alpha\ (RRC(n_4)) + \beta(UE) + \alpha\ (RRC(n_5)) + \beta(gNB\text{-}DU) + \alpha(F1AP(n_6)) + \beta(gNB\text{-}CU)+\beta(MEC) = 6\alpha + 6\beta + 2\beta(MEC) = 4.22 \text{ ms}.$$

The numerical results in [24] regarding the evaluation of MEC latency, use a basic key performance indicator (KPIs): round trip time (RTT). The RTT, shown in Fig. 10, follows the above-mentioned data pattern in time for the flows in Fig. 7 and Fig. 8, and allowing us to develop a cumulative distribution function (CDF) for the empirical data.

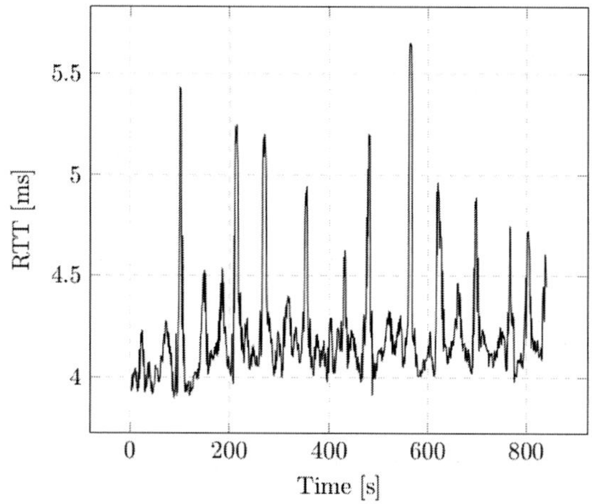

Fig. 10. Empirical data for round trip time

We aim to discover an accurate yet simple theoretical distribution to the data in order to be able to reuse it in a compact manner in cases of MEC-based services simulation/development/tests, and the same question when mobile applications are under focus.

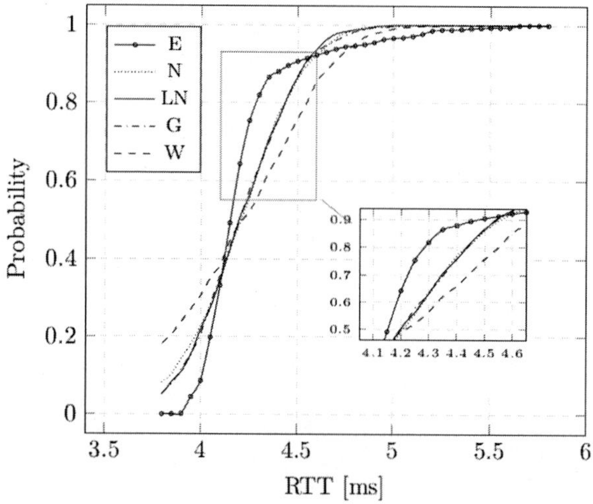

Fig. 11. Empirical cumulative distribution function of round trip time fitted by different theoretical distributions, where: E - empirical distribution; N - Normal (μ = 4.2162016; σ = 0.2704341); LN - LogNormal (μ = 1.43706979; σ = 0.05963488); G - Gamma (α = 268.2829; β = 63.6315); W - Weibull (k = 11.115669; λ = 4.359179).

It seems that there are a reasonable number of candidates for data fitting; we can count on well-known distributions: Normal, LogNormal, Gamma, and Weibull. This list of candidate distributions is very far from exhaustive. However, it is worth mentioning that fat-tail distributions are omitted, because the signaling procedures do not lead to the distributions, such as those characterizing heavily interlinked web traffic.

In the process of obtaining the parameters of the distributions in Fig. 11, the maximum likelihood procedure is used.

The descriptive parameters of candidate distributions, fitted so far, are given in Table 2.

Table 2. Summary of descriptive parameters

	Normal	LogNormal	Gamma	Weibull
Median	4.2109	4.1948	4.2006	4.2140
Mean	4.2038	4.2093	4.2111	4.1731
Std Dev	0.2751	0.2542	0.2615	0.4396
Skewness	0.0195	0.1138	0.2075	-0.711
Kurtosis	2.8985	2.6977	2.9297	3.8959

The descriptive parameters in Table 2 characterize the respective distribution per se, but the selection has to be based on comparison, which requires the running of some tests.

The comparison of fitted distributions to the empirical distribution is shown in Table 2, where the lower the statistical values are the better, i.e. the fit's "discrepancy" is lower.

In Table 3, the abbreviations for the statistics are as follows: Kolmogorov-Smirnov (KS); Cramer von Mises (CvM); and Anderson-Darling (AD). All three types are better than the LogNormal distribution when compared to the other three distributions.

Table 3. Goodness-of-fit statistics

	KS	CvM	AD
Normal	0.21300	12.15371	67.95442
LogNormal	0.19771	10.10511	57.64129
Gamma	0.20298	10.77470	61.02661
Weibull	0.26222	21.45242	112.55178

We double-checked the results shown in Table 3 for our selection procedure by calculating the information criteria where the figures might be found in Table 4.

This has to be done in order to understand the level of information loss that would be introduced by each model. For this purpose, we can use the well-known Akaike's information criterion (AIC) and the Bayesian information criterion (BIC).

Table 4. Information criteria for fitted distributions

	Normal	LogNormal	Gamma	Weibull
AIC	190.84	65.31	105.81	710.04
BIC	200.30	74.78	115.28	719.50

Ranking the results in Table 3 and Table 4, i.e. sorting in ascending order, shows that the LogNormal distribution is the best candidate to fit the empirical data.

Conclusion

5G network infrastructure will exploit network function virtualization to provide improved quality of service and traffic prioritization based on service policies. Multi-access Edge Computing possesses the potential to fulfill the requirements of ultra-low latency, reliability, and availability in mission critical applications.

Proactive dynamic user traffic handling, including QoS control and management, can be realized with a new mobile edge service that provides functionality for user traffic shaping, redirecting, and blocking. In this paper, we have proposed a new mobile edge service that addresses MEC traffic routing requirements. The service is described by means of typical use cases, data model, API definition, and state models.

We have also considered two MEC deployment scenarios appropriate to mission critical applications. One of the scenarios includes deployment of the mobile edge host at the gNB site, where mobile edge applications and services can share the virtualized infrastructure of centralized NG-RAN functions. This scenario features lower latency in the provisioning of radio network information. The other scenario includes deployment of the mobile edge host at the edge site where core network functions are distributed and mobile edge applications and services are running on the same network function virtualization platform as core functions. Based on published empirical data for latency in MEC deployment in a wireless access network, we determined the theoretical distributions and applied these distributions to estimate the latency introduced by the proposed mobile edge service.

Centralized control of user traffic, applied in the vicinity of end-users, is expected to have a positive impact on network performance due to the effectiveness of coordinating and exposing bandwidth information in the radio access network. Uplink user traffic does not need to pass through the network core, as a policy-based decision can be made in the radio access network saving backhaul and transmission resources.

Acknowledgment

This research was conducted under the grant of project DH07/10-2016, funded by the Bulgarian National Science Fund, Ministry of Education and Science.

References

[1] T. Shuminoski, L. Gavrilovska and T. Janevski, "QoS performances of heterogeneous networks with multiple radio access technologies", *IEEE EUROCON 2017. 17th International Conference on Smart Technologies*, Ohrid, 2017, pp.132-137.

[2] R. A.Aljiznawi, N. H. Alkhazaali, S. Q. Jabbar, .D. J. Kadhim, "Quality of Service (QoS) for 5G", *International Journal of Future Computer and Communication*, vol. 6, no. 1, March 2017, pp.27-30.

[3] J. Pedreno-Manresa, P. S. Khodashenas, M. S. Siddiqui and P. Pavon-Marino, "Dynamic QoS/QoE assurance in realistic NFV-enabled 5G Access Networks", *2017 19th International Conference on Transparent Optical Networks (ICTON)*, Girona, 2017, pp.1-4.

[4] A. Reznik, et al. "Cloud RAN and MEC: A Perfect Pairing", ETSI white paper no. 23, February, 2018, pp.1-24.

[5] P. Porambage, J. Okwuibe, M. Liyanage, M. Ylianttila, T. Taleb, "Survey on Multi-Access Edge Computing for Internet of Things Realization", *IEEE Communications Surveys & Tutorials*, 2018, doi: 10.1109/COMST.2018.2849509

[6] R.Bolla, A. Carrega, M. Repetto, G. Robino, "Improving Efficiency of Edge Computing Infrastructures through Orchestration Models", *Computers*, 2018, vol.7 issue 36, pp.1-17.

[7] Q. Ye, J. Li, K. Qu, W. Zhuang, X. S. Shen and X. Li, "End-to-End Quality of Service in 5G Networks: Examining the Effectiveness of a Network Slicing Framework", *IEEE Vehicular Technology Magazine*, vol. 13, no. 2, pp.65-74, June 2018.

[8] Y. Liu, G. Shou, Y. Liu, Y. Hu, Z. Guo, "Performance Evaluation of Integrated Multi-Access Edge Computing and Fiber-Wireless Access Networks", *IEEE Access*, 2018, vol. 6, pp.30269-30-279.

[9] ETSI GS MEC 002. *Mobile Edge Computing (MEC); Technical Requirements*, v1.1.1, 2016

[10] ETSI GS MEC 015. *Mobile Edge Computing (MEC); Bandwidth Management API*, v1.1.1, 2017.

[11] F. Giust, et al. "MEC Deployments in 4G and Evolution Towards 5G, ETSI white paper no. 24, February, 2019, pp.1-24.

[12] ETSI GS MEC 012, *Mobile Edge Computing (MEC); Radio Network Information API*, v1.1.1, 2017.

[13] ETSI GS MEC 013. *Mobile Edge Computing (MEC); UE Identity API*, v1.1.1, 2017

[14] ETSI GS MEC 014. *Mobile Edge Computing (MEC); Location API*, v1.1.1, 2017

[15] E Pencheva, I. Atanasov, "Usage Monitoring Control in Radio Access Network", *23th Conference of Open Innovations Association FRUCT*, Bologna, Italy, 2018, pp.306-314.

[16] 3GPP TS 38.413 Technical Specification Group Radio Access Network; NG-RAN; NG Application Protocol (NGAP), v15.2.0, 2019, Release 15.

[17] 3GPP TS Technical Specification Group Radio Access Network; NR; Radio Resource Control (RRC) protocol specification Release 15, v15.4.0, 2018.

[18] X. J. Chen, R. De Nicola, "Algebraic characteristics of trace and decorated trace equivalences over tree-like structures", *Theoretical Computer Science*, 254, Elsevier, 2001, pp.337-361.

[19] 3GPP TS 38.401 Technical Specification Group Radio Access Network; NG-RAN; Architecture description, Release 15, v15.5.0, 2019.

[20] 3GPP TS 38.473 Technical Specification Group Radio Access Network; NG-RAN, F1 application protocol (F1AP), Release 15, .v15.3.0. 2018.

[21] ITU-R M.2410-0 Minimum requirements related to technical performance for IMT-2020 radio interface(s), 2017

[22] B. Bertenyi et al. "5G NR Radio Interface", *Journal of ICT Standardization*, vol. 6, issue 1 & 2, May 2018, pp.31-58.

[23] D. Soldani, Y. J. Guo, B. Barani, P. Mogensen, C. I. and S. K. Das, "5G for Ultra-Reliable Low-Latency Communications", *IEEE Network*, vol. 32, no. 2, pp.6-7, March-April 2018.

[24] J. Liu, G. Shou, Y. Liu, Y. Hu and Z. Guo. "Performance Evaluation of Integrated Multi-Access Edge Computing and Fiber-Wireless Access Networks", *IEEE Access*, vol. 6, pp.30269-30279, 2018.

CHAPTER 2

INNOVATIVE KNOWLEDGE-BASED SYSTEMS

CHAPTER 2.1

ACCENTED VISUALIZATION FOR AUGMENTED REALITY

ANTON IVASCHENKO[1], PAVEL SITNIKOV[2], OLEG SURNIN[3]

[1]Samara State Technical University
[2]ITMO University
[3]SEC "Open Code"

Abstract. The innovative concept of accented visualization to increase the efficiency of augmented reality (AR) systems for industry is proposed here. The proposed solution is based on the authors' experience of AR implementation. Attention is given to the usability of innovative knowledge-based systems to improve the quality and efficient use of intelligent technologies in digital industry applications. This paper develops the ideas of interactive cognitive visualization based on adaptive construction and virtual consideration of the content of the "real" scenario in the field of view of a person, as well as the viewer's experience, which contains perceptions, points of view, and expected behavior. The proposed solution can be used to identify gaps and operator failures in real time, predict possible operating mistakes, and suggest better procedures by comparing a sequence of actions with the experience of highly qualified decision-makers. Accented visualization technology can be developed further to form a new theory of perceptive co-design devoted to improving the effectiveness of situational monitoring centers through the implementation of intelligent interaction management technologies and taking into account the characteristics of individual selective perception of a situation.

Introduction

Augmented reality (AR) is one of the most promising modern technologies and has the potential to considerably improve the capabilities of interactive user interfaces. Modern AR devices are becoming comparatively cheap and easy to use in most areas of human-computer interaction. Starting with various 3D games that provide computer generated virtual content that is overlaid on a real environment, AR solutions are becoming natural in multiple areas of life including the service sector, manufacturing, the transport industry, medicine, tourism, and education.

Despite the rich functionality and high usability of AR devices, their practical implementation remains a challenge. Such problems relate to both usability and perception. Most users experience difficulties when using head-mounted devices while executing routine manipulations. People treat digital interventions into their private realities in various ways, making it difficult to develop universal solutions for supporting decision-making.

Using the experience of solving these types of issues, the novel concept of accented visualization was developed (Ivaschenko 2019). This concept considers the main feature of AR-based solutions as bringing digital objects into a person's perception of the "real world". Based on an overview of scientific papers and experience in the area of AR solutions development, the model of accented visualization and its software implementation is proposed for the exploitation of technologies drawing on knowledge-based reasoning and intelligent methods of patterns discovery using neural networks, for better adaptation of visual user interfaces.

Implementation and testing have proven the efficiency of the proposed solution, as well as its practical utility. AR-based applications for manual operation control were used to win a specialized contest organized by KAMAZ, the Skolkovo Foundation, and the Foundation for Advanced Studies; these application were nominated as the best industrial solution in the competition organized in August 2018. This paper extends this concept by presenting a new idea, "perceptive co-design", which is proposed here for the first time.

State of the Art

Most modern AR technologies (Navab 2004; Van Krevelen 2007) have powerful and adaptive user interfaces. Various AR devices (goggles, head-mounted displays, or tablets) overlay information or mask the real environment, which provides information to a user. AR technology allows the development of an interactive and context-dependent user interface with possibilities for computer visualization and object recognition in real time.

AR technologies are tailored to adaptive user interfaces (UI) (Julier 2003). AR devices need to present sufficient information at each stage of a user story with no redundant data and maximizing user attraction to important items. As such, an AR-based UI differs from the UI found on personal computers. Additional data and control elements can be placed and presented in an accessible space and in 3D with no limitations of screen or panel size. This data overlaps with real objects; it can hide some of these objects and use others for docking.

When presenting textual information, it should appear at the correct focal length for the user to make it readable without disturbing them. These problems remain unsolved in most practical applications of AR and can be solved by adapting the UI to concrete user cases.

The possibility of AR becoming a ubiquitous user interface in the real world are discussed in Singh (2013), Lee (2012), and Schmalstieg (2005). User interfaces are usually designed with a focus on maximizing usability and user experience. The goal of user interface design is to make a user's interaction as simple and efficient as possible, while accomplishing user goals.

To ensure adequate and stable identification of real objects and items with complex shapes, an algorithm based on a combination of standard technologies of image analysis and neural network-based pattern recognition can be implemented (Goodfellow 2016). Methods for identifying components that have a simple form or consist of elements with a simple shape have been taken from free sources and modified to achieve the required quality and performance.

Contextual data visualization (Holzinger 2014, 2016) sees the combination of several datasets to analyze multiple layers of a biological system at once. The system should interlink all related datasets (e.g., images, text, measured values, and scans) and offer visual analytics to support expertise. This approach is widely used in medical data processing, but it can be easily disseminated in a cyber-physical system. This supports the idea of the maximum effective visualization of complex data for professionals in place of automatic decision-making. In addition to this, the involvement of the decision-maker in the process of data processing and visualization by means of continuous interaction with the system is proposed to help optimize the learning behavior of both humans and algorithms.

The human features of the operator are formalized and captured by a knowledge base (ontology) that can be used to adapt and personalize user interfaces for better understanding and usability. Modern approaches of information-driven and ontology-oriented smart space design and the improvement of interaction involving many participants are presented in Korzun (2017).

An AR system must engage the three overlapping concepts of tracking, registration, and calibration (Schmalstieg 2016). To display virtual objects registered to real objects in 3D space, we must at least know their relative pose, i.e. the position and orientation of the AR display in relation to the real objects. Because AR operates in real-time, pose measurements must be continuously updated, i.e., tracked over time.

In the field of AR, tracking the three-dimensional position of real entities is contrary to the idea of tracking two-dimensional objects in image space, which is common in traditional computer visualization. Calibration is the process of comparing measurements made with two different devices: a reference device and a device to be calibrated. The reference device can be replaced with a known reference value or with a known coordinate system. The objective is to determine the best parameters for using the device to be calibrated so as to deliver measurements in accordance with the known scale. For AR, we need to calibrate the components of the AR system, especially of devices used for tracking.

While tracking means continuously performing measurements, calibration is usually carried out only at discrete times. Registration in AR refers to the alignment of coordinate systems between virtual and real objects (Holloway 1997). Computer graphic elements should align with real-world objects. This requires tracking of the user's head, the camera providing a video background, or both. Obtaining static registration, when the user or the camera is not moving, requires calibration of the tracking system so as to establish a common coordinate system between virtual and real objects. Obtaining dynamic registration, when the user or the camera is moving, requires tracking.

Eye tracking may have the broadest and most profound impact of all human facial signals given that eye tracking can measure human attention. Rendering processes are redesigned only to show the user a portion of what they are looking at in full detail. These techniques operate on the scientifically proven fact that our eyes have a narrow field of view at full resolution and blurring occurs outside the fovea (*The importance of Eye Tracking in AR/VR* 2018).

Preliminary results on the testing of accented visualization are presented in Ivaschenko (2019). Based on the experience of the authors' implementation, the concept of focus and context in an AR-based solution to improve decision-making support in practical applications was developed. As an example, the proposed concept was implemented in a specialized intelligent system for industrial manual operation control. Such a system implements the ideas of Industry 4.0 (Kagermann 2013; Lasi 2014).

AR Case Stories

In addition to the theoretical overview in this section, a description of the authors' practical experience of several projects focusing on the implementation of AR solutions in real-world cases is provided.

The first project was dedicated to developing a customized AR-based interactive guide for a robot design kit. The project was successfully performed for the Abilix robot creative brick series, using tablets and Epson Moverio goggles as AR devices. The resulting AR software solution can be deployed on a tablet or through AR goggles and can be used by children to easily and comfortably investigate how to use the small robot components. Experiments and practical use of AR devices in schools have proven the potential of their application and their interest for students.

Modern robot design kits have powerful capabilities for developing various kinds of construction with programmable behavior, starting from elementary forms and moving on to ever advancing levels of complexity. These design kits have a lot of potential for education and for training a child's creative, analytic, and practical abilities. As a result, they are becoming widely implemented in middle schools.

The technical challenges of implementing an AR guide for a robot design kit concern the low performance and quality of image recognition algorithms capable of functioning in real time. The main problem of AR implementation concerns the stable and reliable recognition and identification of objects in view. For example, one of the main challenges of AR user guides is the need to identify a currently required object, attract the user's attention, and provide complete and comprehensive annotation.

AR-based solutions are dedicated to supporting the process of finding small pieces (details) easily and being comfortable to use. A special software package has been developed (see Fig. 1), which can identify various components using a combination of existing image recognition techniques. Visualization of the results of the identification process can be performed through hints (see Fig. 2) and selection (see Fig. 3).

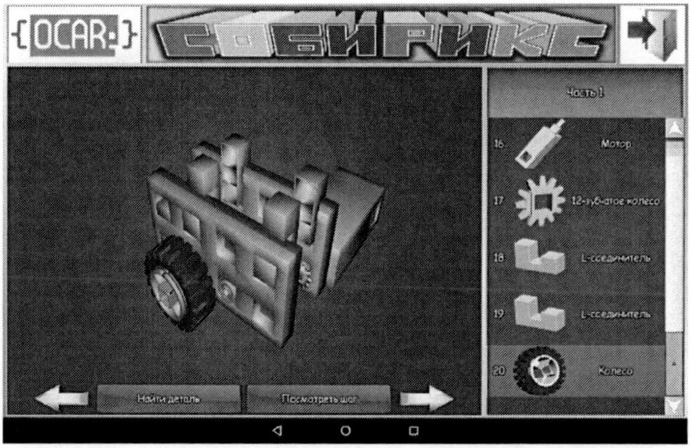

Figure 1. Robot design navigator

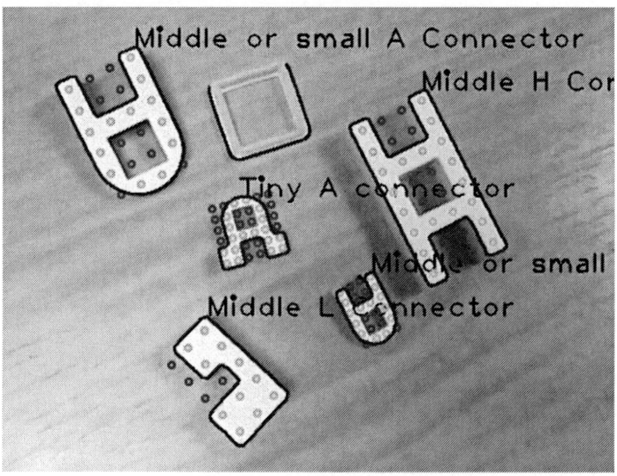

Figure 2. Design kit component identification

Intelligent systems for manual operation control are close, in terms of use, to the functionality of interactive user guides that also implement AR for decision-making support at certain stages of operational processes. Intelligent software provides image recognition of objects and their matching to a corresponding description in a knowledge base. Video panels or AR goggles are used to present the corresponding contextual information to an operator.

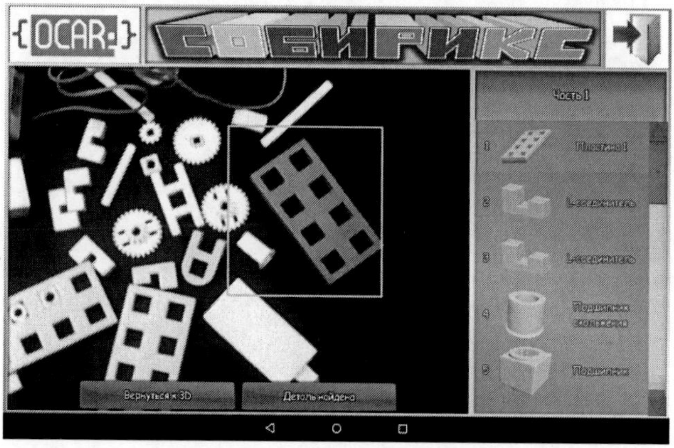

Figure 3. Identification algorithm result

A particular truck component, the radial compressor (turbo machinery), was used as an object to test recognition. The object identification process is illustrated in Fig. 4. The parts of the required object are contextually identified (see Fig. 5) and highlighted according to the production process (see Fig. 6). The primary condition of the experiment was the use of an ordinary cheap webcam for the task of recognition and tracking.

The software controls the correctness of the product assembly. The initial data describes the assembly process with typical errors, the digital model, and the assembly video. The main parts of the product include: the compressor wheel; the cover; the compressor housing; the bearing housing; the turbine housing; the turbine screen; the oil-drain adapter; the rotor; the bearing housing with a rotor; and the assembled radial compressor.

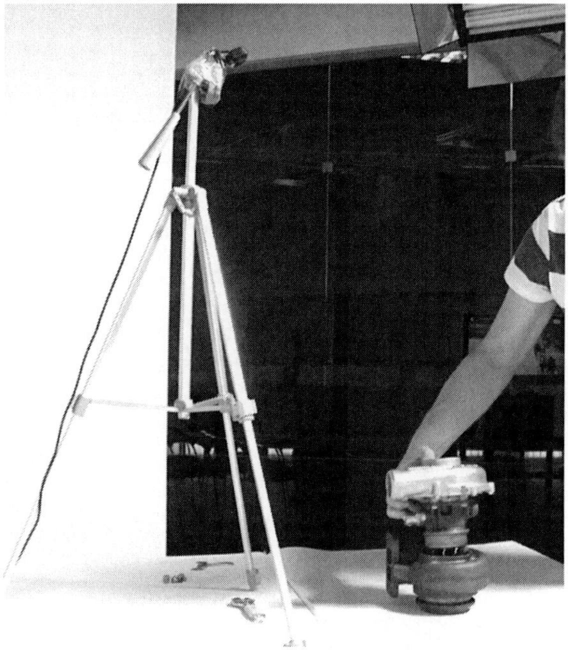

Figure 4. Intelligent manual operation control

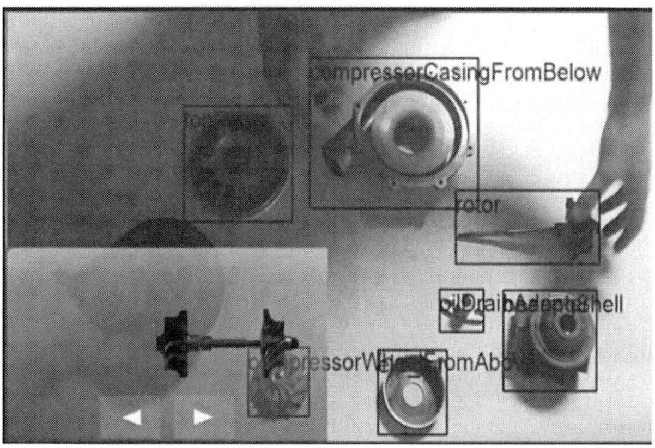

Figure 5. Intelligent object identification

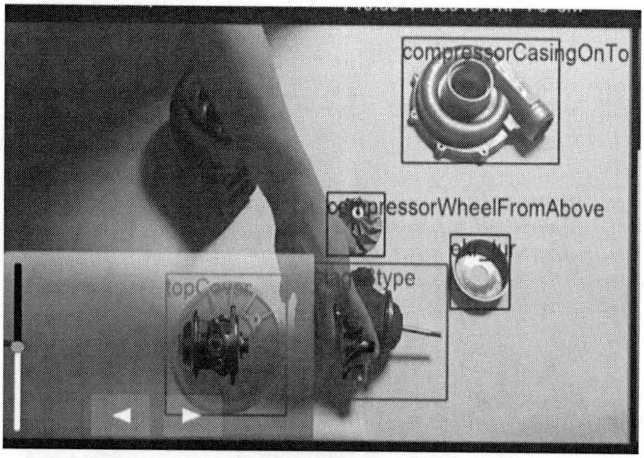

Figure 6. Intelligent manual operation control. Relevant object highlighted

Practical deployment can be done either by using a specifically developed head-mounted device (see Fig. 7) or AR goggles (see Fig. 8).

Figure 7. Intelligent manual operation control. Head-mounted device

Figure 8. AR interactive guide

The solution requirements include the following:

- it should visually (in AR mode) present a step-by-step demonstration of the process for assembling the layout in accordance with the procedure described in the technological process for the product;
- it should display an error if the product is not assembled in accordance with the proper technological process;
- it should display an error if during assembly of the product typical errors are listed in the process.

However, problems of using AR goggles in practice include the necessity of personalizing them and ensuring their extended usability and comfort. In this respect, other possible ways of deploying a system using cheap and widespread equipment have been explored. A number of video cameras are used to track operations according to production processes and identifying objects to be manipulated in reality.

The overall solution is used to identify gaps and operator failures in real time, predict possible operating mistakes, and suggest better procedures based on comparing the sequence of actions taken to the experience of highly qualified operators.

Accented Visualization Model

In this section, a combined formal model for contextual visualization of overlay data using an AR device is proposed. This model was originally developed for maintenance manuals of technical equipment and is based on the experience of implementing AR and VR projects; it can be extended and used in industrial applications.

The notion of accentuated visualization is based on the adaptive construction and virtual consideration of the content of not only the current real scene in a person's field of view, but also the viewer's experience containing perceptions, points of view, and expected behavior. This system allows the solution of the situation, as performed by different users with different levels of education, knowledge, and skill, to be seen in a unified way.

Consequently, virtual and real objects are integrated together generating immersive sensations that are perceived as a natural part of the environment. Accented visualization formalizes and controls this process, taking into account individual user behavior and adapting the user interface accordingly.

Let us consider the scene s_j, where $j = 1..N^s$ is the scene number containing various real objects $w_{i,j}$, $i = 1..N_j^w$ and their corresponding virtual entities $q_{i,j,l}$, $l = 1..N_{i,j}^q$:

$$s_j = \left\{w_{i,j}, q_{i,j,l}\right\}.$$

Objects are considered to be independent in the view of various features of visualization in various scenes. Objects can have different displays in different scenes, but refer to the real world.

These objects are considered equal. It is possible to describe this equality using the compliance matrix for all scenes:

$$C\left(w_{i_1,j_1}, w_{i_2,j_2}\right) = \begin{cases} 1, w_{i_1,j_1} \equiv w_{i_2,j_2}, \\ 0, elswise. \end{cases}$$

The viewer's attention, at each point in time, is given to many objects (one or more). Each center of an object can be described by an event represented as a Boolean variable:

$$v_{i,j,k} = v\left(w_{i,j}, t_{i,j,k}\right) \in \{0,1\}.$$

This focus may end or require action. In this case, the sequence of actions can be considered as a scenario. For example, the service process for a particular device can be described by a scenario of actions organized in sequence.

The working process is formulated in a standard scenario:

$$c_{j,m} = \left\{e_{i,j,m,n}\right\},$$

where $e_{i,j,m,n} = e\left(w_{i,j}, d_{i,j,m,n}, t_{i,j,m,n}, \Delta t_{i,j,m,n}\right) \in \{0,1\}$.

Events $e_{i,j,m,n}$ represent the facts that actions $d_{i,j,m,n}$ referring to the objects $w_{i,j}$ have to be done at lap-time $t_{i,j,m,n} \pm t_{i,j,m,n}/2$.

An effective process requires the correspondence of $e_{i,j,m,n}$ and $v_{i,j,k}$:

$$M\left(e_{i_1,j,m,n} \cdot v_{i_2,j,k}\right) = e_{i_1,j,m,n} \cdot v_{i_2,j,k} \cdot \left[w_{i_1,j_1} \equiv w_{i_2,j_2}\right] \cdot$$
$$\cdot \left[w_{i_1,j_1} \equiv w_{i_2,j_2}\right] \cdot \left[t_{i,j,k} = t_{i_1,j,m,n} \pm \Delta t_{i_1,j,m,n}\right].$$

AR implementation in this case should support maximum correspondence of $v_{i,j,k}$ to $e_{i,j,m,n}$ resulting in prompt contextual data visualization according to the viewer's current focus. It can be formalized as generating a set of focus attractors in textual, marked-up, or highlighted form:

$$\mathrm{Q}_{\mathrm{i,m}} = \left\{q_{i,j,l}\right\}, q_{i,j,l} = q\left(w_{i,j}, t_{i,j,l}\right) \in \{0,1\}.$$

to meet the following objective:

$$I\left(s_j, c_{j,m}\right) = \sum_{n=1}^{N_{j,m}^{c}} e_{i,j,m,n} \cdot$$
$$\cdot\left(1 - \left[\sum_{k=1}^{N_j^v} \sum_{l=1}^{N_j^q} \sum_{i=1}^{N_j^w} \left(M\left(e_{i,j,m,n} \cdot v_{i,j,k}\right) \vee q_{i,j,l}\right) > 0\right]\right) = 0$$

This means that, in order to provide a standard AR scenario, an interactive guide should produce virtual enterprises along with a context to define the data in moments when the viewer's attention needs to be drawn to specific objects.

Solution Approach

The concept of accented visualization was developed according to the methods of contextual data visualization and interactive computer-human interaction. The system architecture is presented in Fig. 9.

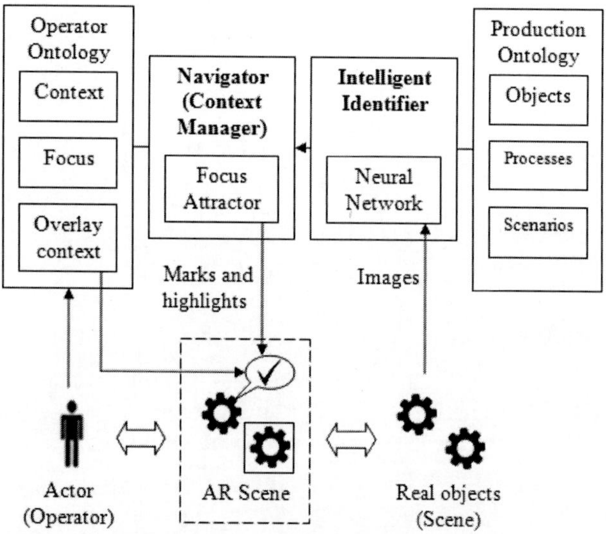

Figure 9. Accented visualization solution

The concept is based on the formalization <focus, context, and overlay context> for each actor in the operator ontology. The context uses a set of concepts that describe the current situation and the background that determines the decision. The focus is a concrete object processed at a certain moment in time. The overlay context includes virtual entities (textual items, marks, or highlights) attracting the user's attention to objects in the scene when needed. Such fragmentation allows the introduction of a control loop and the correct focus is stimulated in real time according to the context.

The navigator module (context manager) implements coordination of the user's focus and is based on intelligent analysis of production processes. The user's focus is captured in the form of event chains and compared to typical scenarios. The system tracks user attention and adapts additional data from the overlay context introduced into the virtual scene according to current needs.

The identifier module implements intelligent pattern recognition using, for example, neural networks. It is a basic component of the whole solution and provides identification of all the objects in view and navigator data support.

The production ontology involves a knowledge base that contains the description of production processes, technologies, and necessary equipment, as well as the critical features or characteristics of each object. Their processing rules are taken from production processes and coordinated with other scene objects and steps from the operating scenarios. Eye-tracking software and hardware is used to identify possible gaps in the viewer's perception if the required attention is not given to particular scene objects at the required time.

Despite the variety of shapes, the details of most of them can be identified using the same approach. Some components look similar, while some others differ only in size or color. In addition to this, the AR device should be capable of identifying overlapped details with a lack of required visual data. To overcome this difficulty and increase the efficiency of the AR interactive user guide, a new approach based on accented visualization was developed considering the current context.

Implementation of the accented visualization approach considers that the combination of focus and context can be presented in the form of two layers: the identifier directly for intelligent pattern recognition and the navigator acting as a wrapper for the identifier. The navigator can be used for pre-processing of input data, reducing obvious mistakes and interference. It can be used for post-processing of identification results to improve their interpretation in support of decision-making.

The context at the second layer can provide better targeting of the intelligent algorithm, or it can help choose the most appropriate algorithm from a list. In this perspective, it can be implemented using formal logic predefined by an expert (as in the example given above). A special visual tool to describe this logic in the form of a decision-making tree or semantic network can be proposed as a part of the ontology. This tool can have a special user interface for several experts and later be used for configuration of the system.

The navigator can also implement its own neural network as an alternative; it is trained in a different way to the first layer and therefore becomes capable of producing additional logic of data pro-processing and interpreting. Input in this layer will take the distribution of probabilities of object identification from the first layer and determine the context in the form of, for example, the stage of the process.

Possible issues of implementation of the proposed approach relate to its high dependency on the first layer, which remains the main source of information and a basic identification tool; as well as the complexity of decision-making trees or semantic networks in the second layer in the case of processing multiple objects, stages, cases, and processes. Nevertheless, this approach remains effective for putting intelligent technology into practice and increasing adequacy and sufficiency in industrial applications.

Implementation and Testing

Implementation was divided into three stages. The task was to detect objects in the video stream, regardless of their location (objects could be on the floor, on the table, near foreign objects, or next to other objects). Furthermore, the object may not be in the frame at all.

At the first stage, the task was to automatically determine the coordinates of parts of a product and an assembled product on video images for a test sample. To provide adequate and stable identification of objects with complex shapes, we developed an algorithm based on a combination of standard computer visualization methods and neural network-based pattern recognition.

Methods of identifying components that have simple geometric shapes and/or consist of elements with simple geometric shapes were taken from free sources and modified to ensure the required quality and performance. The proposed identification algorithm includes the following steps:

1. The outer contour is identified for each object.
2. The contour is drawn around the rectangle with the smallest possible area.
3. The ratio of the length and width of the rectangle is calculated and this must correspond to the analogous value at the contour of the expected part.
4. A dot grid is applied to the rectangle.
5. A map is drawn describing which points are inside the contour (belong to a figure) and which are not.
6. The compiled map is compared with a pre-prepared map.
7. In case of coincidence, the parts are identified.

The task at the second stage was how to realize the loading of multiple video files and datasets, after which the resulting output is in the form of a .CSV file containing a table with the coordinates of recognized product elements. Each video file was processed independently of other video files in the sample, i.e. no additional information about the other video files was used during processing. A log of the real-time solution of the problem was also kept. The average processing time for a single video file did not exceed the playback time of the video file itself.

At the third stage, a real-time solution was launched with a video stream received from a webcam mounted above the workspace. Software functionality at the third stage includes the following features:

- scene object identification based on image analysis;
- contextual description of the object in view;
- search and highlighting of the object;
- operating scenario processing, tracking, and control;
- complex devices analysis, including component identification in partial view and generation of tips for assembly;
- user attention identification and contextual add-on generation according to the principles of accented visualization.

In order to provide the required performance, in addition to the described algorithm, the approach of accented visualization was implemented. User focus coordination was introduced and based on analysis of the design process. The system tracks user attention and adapts additional data according to current context and need.

The problem of AR implementation concerns the stable and reliable recognition and identification of objects in view. For example, one of the main challenges of AR user guides is the necessity of identifying the object required at the current moment, attracting the user's attention, and giving complete and comprehensive annotation. To provide such features the technology should be capable of processing a substantial number of images in real time.

One can see that no extra limitations for the place of operation and lighting (e.g. a green background) are required. Simple web cameras can be used to track and capture objects and the video quality is good enough for the most up-to-date models. The use of a minimum of two cameras is recommended to reduce lighting defects caused by occultation and blur. The quality of lighting turns out to be not so critical, making this technology suitable for deployment in real enterprises. Considerable lighting changes require additional calibration.

In practice, intelligent image recognition and object identification is performed using standard neural networks. Several alternative libraries, including Tensorflow (which is the fastest), Keras, Theano (no longer receiving developer support), and Deeplearning4j (supports Java) were considered. Tensorflow was chosen for implementation because, in addition to its high productivity, it is distributed under the open license Apache 2.0 and provides access with Python, C++, Java, Haskell, Go, Swift API, supports Linux, Windows, macOS, iOS, Android; supports Google and cloud computing; and is popular with developers.

Of the possible neural network topologies (AlexNet, VGG16, GoogLeNet/Inception, etc.) ResNet-50, a residual network powered by Microsoft, was chosen. ResNet-50 is a convolution neural network that has been trained on more than a million images from the ImageNet database. It goes to a depth of 50 layers with direct links between neurons located by level.

Despite the variety of object shapes, in practice most industrial components look fairly similar. In addition, the solution should be capable of identifying overlapped details with a lack of visual data. For example, the assembly unit has two details and is a combined bearing case and adapter (see Fig 6). The adapter does not add much to the unit's identity (providing only 10-15 % of the overall picture space), so the two pictures of the bearing case and the assembly unit appear very similar. As such, the standard intelligent identification algorithm mixes them up.

Problems like this can be solved using the proposed concept of accented visualization. In this approach, the intelligent identification algorithm is wrapped by knowledge driven logic, which coordinates its target on the basis of a correlation between the user's focus and the context. To support context driven decision-making, a table of objects with possible distributions and probabilities of occurrence was built. This table forms a decision-making tree determining the sequence of stages and requirements for a list of details to be presented at each stage. As such, the current stage determines the priority of object identification.

For the above example, a rule can be specified that in the case of the combination of a bearing case and assembly, if an adapter is identified to an analyzed object with a probability of 0.7 then the object is identified as a bearing case. This logic corresponds to the 5S methodology of workplace organization, which requires no foreign objects exterior objects to appear in scene.

Its implementation in the real case of a radial compressor assembly and identification of its details considering the requirements of the production process are given in Fig. 10. The functions represent the dependency of a number of faults in the identification stage.

The faults were calculated in the following way. The video of the identification procedure was recorded (with a duration time of 9.13 min) and analyzed by an expert. Every 10th frame was used for analysis and the object identified by the system was compared to the object required for identification according to the production process. The total number of steps was 1,271; this is shown in the figure. Each step corresponds to a certain frame of the scene.

Two types of identification were investigated: with context and without context. The case without context, where the neural network was implemented only for the intelligent identification of objects in view is illustrated by the upper function. The total number of faults was 802, putting it in the range of 63.1 %. The second case considered the context and saw 86 faults, giving it a fault ratio of 6.8 %.

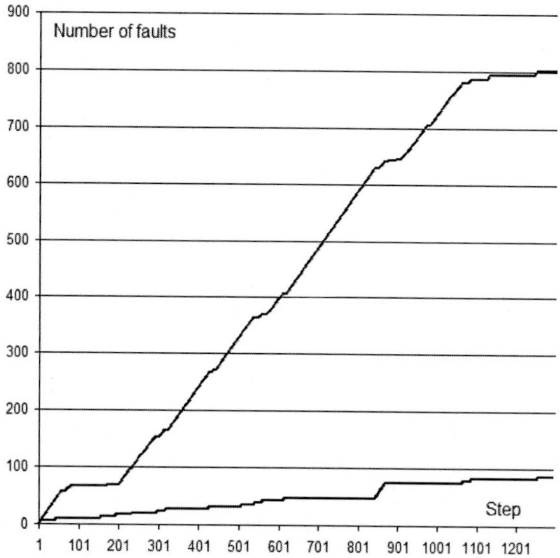

Figure 10. Accumulation of identification faults (errors): considering context (lower graph) and no context (upper graph)

This example shows that the successful recognition rate of objects through identification with accented visualization was 93.2 %, compared to the standard technology, which gave a value of 36.9 %.

This intelligent system for manual operation control with identification won a special contest organized by KAMAZ, the Skolkovo Foundation and the Foundation for Advanced Studies; it was nominated as the best industrial solution in August 2018.

The Concept of Perceptive Co-design

Accented visualization technology can be further developed leading to a new theory of perceptive co-design. The new organizational approach to user interfaces sees an increase in the effectiveness of situational monitoring centers through the implementation of intelligent interaction management technologies, taking into account the characteristics of selective perception of a situation by decision-makers.

New interaction conditions between participants in digital economic relations have significant influence on the cognitive features of perception, through which the user interfaces of interactive decision support systems are oriented. This emphasizes the relevance of developing such methods and the means of organizing situational centers, taking into account the need for cooperative interaction between decision-makers, which occurs predominantly in a virtual environment, and the changing dynamics of real-time situations.

Nowadays, the theoretical and practical problems of situational management; project and financial planning; risk management; and the organizational management of complex systems are becoming ever more widely developed. However, there is no comprehensive solution to issues related to managing the interaction of decision-makers in new conditions substantially involving digital forms of interaction. In this regard, the scientific and technical problem relies on the intellectualization of situational centers under condition of increasing information flow, high dynamics of change, and the combined influence of time and human factors.

Theoretical novelty is determined by the subject of research, which is here seen in the impact of aggregation and decomposition of information to decision-makers, in terms of the effectiveness of cooperative interaction in a single information space of a virtual organization or social network.

The approach is illustrated in Fig. 11. The proposed theory summarizes the research results and contains the following statements.

Figure 11. The concept of perceptive co-design

A new concept of perceptive co-design of data environments in situational centers is introduced, taking into account the peculiarities of "big data" perception and the virtual interaction of participants in digital economic relations. The model of accented visualization and technology for monitoring indicators of the socio-economic system can be applied in distributed organizational structures, characterized by the possible decomposition and aggregation of indicators, taking into account individual and collective features of the perception of decision-makers.

Methods of semantic-statistical analysis and cognitive visualization of big data, based on knowledge and the development of complex organizational, technical and socio-economic systems, production cooperatives, logistic centers, and social networks are proposed.

An original software enterprise content management platform was developed, expanding the possibilities of document circulation by means of semantic-statistical analysis and allowing the implementation of the concept of perceptive co-design in monitoring systems. This effects regional portals of state and municipal services; social network analysis systems; and social and economic situational centers.

The theory of perceptive co-design reveals the features of perception and information interaction among individuals who make cooperative decisions according to their analysis of big data from a single enterprise's information space, including the logistics of a supply chain or social network.

Analysis of these features allows the construction of cognitive interfaces for control systems in social and economic systems using modern virtual and augmented reality technologies. The required effectiveness of situational centers, in terms of selecting a necessary and sufficient set of information to support decision-making in real time, is ensured.

The building of monitoring tools based on the accented visualization concept was implemented in the BI/big data analytical platform, which, unlike analogue systems, provides cognitive visualization of indicators using large amounts of semi-structured data for real-time cooperative interaction between decision-makers.

Based on the proposed methods and tools, a software platform for enterprise content management system intelligence was developed by the software engineering company Open Codc, allows the extension of the capabilities of electronic document management with options for creating and controlling tasks through new semantic searches and cognitive visualization to support decision-making.

The platform contains the following components.

Knowledge base: a data warehouse at the conceptual and logical levels, in which data is stored as a semantic network of entities, relationships, and output rules. Object or relational DBMS, file storage, or a NoSQL database can be used as physical storage.

Integration bus: an infrastructure software solution that provides inter-module interaction within the service-oriented architecture, intersystem interaction, and integration with external solutions.

Semantic search engine: a complex of methods forming a full-text index for the performance of a productive search, including searching by words and word combinations, word forms, synonyms, and meaning; as well as whether the search query is not clearly defined or incomplete.

User interface designer: a component that allows the design of custom windowed web interfaces (object cards, trees, lists, search forms) without writing any code at the configurational level.

Analytical platform: a component that organizes mathematical calculations in order to obtain aggregated analytical data, including aggregated states with grouping, time, or spatial series; OLAP-cubes; linear regression; and factor and correlation analysis.

Accented visualization toolset: the module presents information to the user in the form of infographics, with reference to cartographic and geo-information, in the form of 3D models, as well as through augmented reality.

Intellectual scheduler: a complex of methods and algorithms (analytical, students, agent) to perform adaptive production planning, project management, and solve logistics problems.

Object-relational database designer: a component to create and maintain a database structure at the object level, without the need to write program code for a specific DBMS.

Document flow and archives: components for organizing the electronic document management of an enterprise; creating and monitoring the progress of tasks and assignments; and creating archives of organizational, administrative, and project documentation with the ability to store not only indexes, but also multimedia information in large volumes.

Big data processing service: methods, algorithms, and infrastructure solutions for processing big data, including methods of image processing, computer visualization, text recognition, text analysis, textual understanding, and the organization of parallel computing.

Conclusion

Using the research presented here, a new solution to improve the efficiency and usability of augmented reality systems to fulfil industrial needs was developed. The ideas of accented visualization and perceptive co-design are proposed as solution for improving the effectiveness of situational monitoring centers through the implementation of intelligent innovative knowledge-based systems, taking into account the characteristics of selective perception of any situation by an individual.

Bibliography

Ivaschenko, Anton, Michael Milutkin and Pavel Sitnikov. 2019. "Industrial application of accented visualization based on Augmented Reality". *Proceedings of the FRUCT'24*, FRUCT Oy, Finland (2019): 123-129

Navab, Nassir. 2004. "Developing killer apps for industrial Augmented Reality". *IEEE Computer Graphics and Applications* 24 (3): 16-20

Van Krevelen, Rick. 2007. *Augmented Reality: technologies, applications, and limitations*. Vrije Universiteit Amsterdam, Department of Computer Science. DOI: 10.13140/RG.2.1.1874.7929.

Julier, Simon, Mark A. Livingston, J. Edward Swan II, Yohan Baillot and Dennis Brown. 2003. "Adaptive user interfaces in Augmented Reality". *Proceedings of ISMAR'03*, Tokyo, Japan

Singh, Mona, and Munindar P. Singh. 2013. "Augmented Reality interfaces". *IEEE Internet Computing* 17: 66-70

Lee, Kangdon. 2012. "Augmented Reality in education and training". *TechTrends* 56 (2): 13-21

Korzun, Dmitry G., Sergey I. Balandin, Alexey M. Kashevnik, Alexander V. Smirnov and Andrei V. Gurtov. "Smart spaces-based application development: M3 architecture, design principles, use cases, and evaluation". *International Journal of Embedded and Real-Time Communication Systems* (*IJERTCS*) 8 (2017): 2, doi:10.4018/IJERTCS. 2017070104

Schmalstieg, Dieter, and Gerhard Reitmayr. 2005. "The world as a user interface: Augmented Reality for ubiquitous computing". *Location Based Services and TeleCartography* (2005): 369-391. DOI: 10.1007/ 978-3-540-36728-4_28

Goodfellow, Ian, Yoshua Bengio, Aaron Courville and Y. Bengio. 2016. *Deep learning, vol. 1.* MIT press Cambridge 925, 773 p.

Holzinger, Andreas. 2014. “Extravaganza tutorial on hot ideas for interactive knowledge discovery and data mining in biomedical informatics”. *Lecture Notes in Computer Science, 8609*, (2014): 502-515

Holzinger, Andreas. 2016. “Interactive machine learning for health informatics: when do we need the human-in-the-loop?” *Brain Informatics,* 3 (2): 119-131

Schmalstieg, Dieter, and Tobias Hollerer. 2016. *Augmented Reality: Principles and Practice*, Pearson Education

Holloway, R.L. 1997. “Registration error analysis for augmented reality. Presence”, *Teleoperators and Virtual Environments* 6 (4): 413-432

The importance of Eye Tracking in AR/VR. Moor Insights & Strategy (2018)

Kagermann, Henning, Wolfgang Wahlster, and Johannes Helbig. 2013. *Recommendations for implementing the strategic initiative Industrie 4.0*. Final report of the Industrie 4.0 Working Group (2013). 82 p.

Lasi, Heiner, Peter Fettke, Thomas Feld and Michael Hoffmann. 2014. “Industry 4.0”. *Business & Information Systems Engineering* 4(6): 239-242

CHAPTER 2.2

ESTIMATION OF RELIABILITY FOR AGROSMART DATA IN KAZAKHSTAN

AIMAN ISKAKOVA, JAMILYA JUMABAYEVA, SAULE BURGUMBAYEVA, AIGUL BAKIRBEKOVA
Gumilyov Eurasian National University, Kazakhstan
ayman.astana@gmail.com

Abstract. The main concept of this chapter is the development of data on reliability for AgroSmart products (crop management software) in Kazakhstan. The reliability function for a new model of the distribution of probabilities of data distortion of AgroSmart products has been studied. Statistical methods have been used to analyze data from AgroSmart products in Kazakhstan with the value of the reliability function estimated in the range of 45 to 100 %. This means that AgroSmart products show no data distortions.

Introduction

This chapter builds on previous work in Iskakova, Jumabayeva, Burgumbayeva and Bakirbekova (2019). In this chapter, a comparative analysis is made to achieve unbiased estimates. The results of this estimation form the main aspect of this work, which consists of uncovering the reliability of data on AgroSmart products.

Khadys, Sikhimbayeva, and Bozhkarauly (2018) and Rustemov, Abikayeva, Rakhimova, Omarkozhayeva, and Temirova (2018) have previously noted that the basis of the Argosmart application lies in the idea of the existence of perturbations within an agricultural field. To assess these perturbations, the global positioning system (GPS) is used, consisting of a subsystem of spacecraft (satellites: SC); a subsystem of control; and a subsystem of consumers.

A feature of GPS signals is that they have very low power and low noise at the surface of the Earth (Zhou, Wang, He, & Wu, 2016). This is due to the low power of the signal emitted from GPS satellites (about 60 W) and the long range of the signal (approximately 20,000 km). Using broadband signals, through correlation processing, signals from this network of spacecraft can be detected and captured.

Iskakova, Jumabayeva, Burgumbayeva, and Bakirbekova (2019) used results from Ayman (2014, 2016) to construct a new model of the probabilistic distribution of data distortions, obtained from the AgroSmart software application. An, Seok, and Park (2016); Jacquesab, Béchetc, and Kerschena (2017); and Messerschmitt (2012) all noted distortions caused by external factors. These causes include: satellite GPS errors (ephemeris and satellite clocks); Earth atmosphere errors (in the ionosphere and troposphere); and user receiver errors (frequency drift, phase drift pseudo-noise sequence, signal detection time).

The main result of Ayman (2016) was the unbiased estimation, with good asymptotic properties, for the probability of distortions obtained by using AgroSmart products.

Nonetheless, there remain many problems with the data. For example, it is necessary to form estimations of reliability for AgroSmart data. This can be done through a comparison of the risks of unbiased estimation with the various aforementioned external factors.

Multivariate Discrete Distribution Generated by the Sums of Random Matrices and the Reliability Function

We consider a probabilistic model of distortion processes in the AgroSmart data.

Let s be the number of possessions from possible m possessions, which show distortion in the AgroSmart data. If the elements of the vector

$$\mathbf{x} = (x_1, x_2, \dots, x_m)$$

present indicators of AgroSmart data distortions for m positions and for every i ($I = 1, 2, \dots, m$)

$$x_i = \begin{cases} 1 & \text{if it has distortion for } i-\text{th position,} \\ 0 & \text{otherway} \end{cases}$$

then the value of s can be presented as the sum of the vector elements $\mathbf{x}$:

$$s = \sum_{i=1}^{m} x_i.$$

The vector **x** is the realization of a random vector

$$\mathbf{X} = (X_1, X_2, ..., X_m)$$

and for each $i = 1, 2, \ldots, m$

$$P(X_i = 1) = p_i$$

or

$$P(X_i = 0) = 1 - p_i.$$

Therefore it is obvious that

$$P(X_i = k) = p_i^k (1 - p_i)^{1-k}$$

where k is equal to 1 or 0. So, if all elements of the vector **X** are random variables, then the probability that a random vector **X** takes on the volume of vector **x** is defined by the formula

$$P(\mathbf{X} = \mathbf{x}) = \prod_{i=1}^{m} p_i^{x_i} (1 - p_i)^{1-x_i}. \tag{1}$$

The value of s can be obtained V_s ways where

$$V_s = \binom{m}{s}$$

and the probability that a random variable S takes on the volume of a value of s is defined by the formula

$$P(S = s) = \sum_{s=\sum_{k=1}^{m} x_k} \prod_{i=1}^{m} p_i^{x_i} (1 - p_i)^{1-x_i}. \tag{2}$$

In other words, the V_s value is a number of partitions of the number s by elements of vectors **X**. Hence, the reliability functions of distributions (1) and (2) have the forms

$$R(\mathbf{X}=\mathbf{x}) = 1 - \sum_{(t_1,\dots,t_m)\left\{\begin{array}{l}\notin \forall i(i=1,..,m) t_i \geq x_i, \\ \text{if } \exists i: t_i = x_i \quad \text{then } \exists j: i \neq j \text{ and } t_j < x_j\end{array}\right.} \prod_{i=1}^{m} p_i^{t_i}(1-p_i)^{1-t_i} \tag{3}$$

and

$$R(S=s) = 1 - \sum_{s=\sum_{k=1}^{m} x_k} \sum_{(t_1,\dots,t_m)\left\{\begin{array}{l}\notin \forall i(i=1,..,m) t_i \geq x_i, \\ \text{if } \exists i: t_i = x_i \quad \text{then } \exists j: i \neq j \text{ and } t_j < x_j\end{array}\right.} \prod_{i=1}^{m} p_i^{t_i}(1-p_i)^{1-t_i}. \tag{4}$$

Estimation of the Distribution of Distortion in AgroSmart Product Data

In practice, the vector

$$\mathbf{p} = (p_1, \dots, p_m)$$

is unknown. Therefore, formulas (1) to (3) are not used in the realization and construction of estimates of probabilities; the reliability functions are the problem to be addressed.

If we assume that we have n data of distortion

$$\mathbf{y}_1, \mathbf{y}_2, \dots, \mathbf{y}_n \tag{5}$$

while using AgroSmart products. For any k ($k = 1, 2, \dots, n$) $\mathbf{y}_k$ from (5) is a vector

$$\mathbf{y}_k = \left(y_{1_{ki}}, y_{2_k}, \dots, y_{m_k}\right)$$

which has the distribution (1).

We denote the statistics

$$s_1, s_2, \dots, s_n \tag{6}$$

where for $k = 1,.., n$

$$s_k = \sum_{i=1}^{m} y_{i_k}.$$

Consider the statistic

$$\mathbf{z} = \sum_{k=1}^{n} \mathbf{y}_k \tag{7}$$

which is a sufficient statistical total. Using Rao-Blackwell-Kolmogorov we can derive unbiased estimations with uniformly minimal risk for probabilities (1) and (2) corresponding to

$$P^*(\mathbf{X} = \mathbf{x}) = \frac{1}{n^m} \prod_{i=1}^{m} \binom{z_i}{x_i} \binom{n - z_i}{1 - x_i} \tag{8}$$

and

$$P^*(S = s) = \frac{1}{n^m} \sum_{s=\sum_{k=1}^{m} x_k} \prod_{i=1}^{m} \binom{z_i}{x_i} \binom{n - z_i}{1 - x_i}. \tag{9}$$

As such, the unbiased estimator with uniformly minimal risk for the reliability function (4) is

$$R(S = s) = 1 - \frac{1}{n^m} \sum_{0 \le t < s} \sum_{t=\sum_{k=1}^{m} x_k} \prod_{i=1}^{m} \binom{z_i}{x_i} \binom{n - z_i}{1 - x_i}. \tag{10}$$

Comparison of the Risks Inherent in the Unbiased Estimation of the Distribution of Distortions in AgroSmart Data

Ziskind and Wax (1988) suggest that for distribution (2), a maximum likelihood estimation exists if all the elements of the observations or realizations of samples have no more than one part per partition. As is known in practice, often some element of the implementation of the sample has more than one partition. As such, the method for determining the likelihood estimation is not actually applicable to this model.

Consequently, it is obvious that all estimates from a set of unbiased estimates, obtained using (7) and (8), should be prioritized. We can then compare these unbiased estimates.

Let j_1, j_2 1,…, n

$$e_{j_1,j_2} = \begin{cases} 1 & \text{if } s_{j_1} = s_{j_2}, \\ 0 & \text{if } s_{j_1} \neq s_{j_2}. \end{cases}$$

We define the empirical estimation of a probability P(**U** = **u**), formed in (6), as

$$c(S = s) = \begin{cases} \frac{1}{n}\sum_{j_2=1}^{n} e_{j_1,j_2} & \text{if } \exists s_{j_1} : s = s_{j_1}, \\ 0 & \text{otherwise.} \end{cases}$$

Following the results presented in Inglis and Gilmore (2014, 148) it is almost certain that for $n \to \infty$

$$\prod_{i=1}^{n} \left| c(S = s_i) - P_g^*(s_i) \right| = \min_{j=1,\ldots,\mu} \prod_{i=1}^{n} \left| c(\mathbf{U} = \mathbf{x}_i) - P_j^*(s_i) \right|.$$

Similarly, if the empirical estimation for the reliability function (7) is

$$rc(S = s) = \begin{cases} 1 - \frac{1}{n}\sum_{s^1=t}\sum_{j_2=1}^{n} e_{j_1,j_2}, & \text{if } \exists s_{j_1} : s = s_{j_1}, \\ 1, & \text{otherwise,} \end{cases}$$

then it is almost certain that for $n \to \infty$

$$\prod_{i=1}^{n} \left| rc(\mathbf{U} = \mathbf{x}_i) - R_g^*(\mathbf{x}_i, \mathbf{z}_g) \right| = \min_{j=1,\ldots,\mu} \prod_{i=1}^{n} \left| rc(\mathbf{U} = \mathbf{x}_i) - R_j^*(\mathbf{x}_i, \mathbf{z}_j) \right|.$$

Let $j = 1,\ldots, \mu$

$$\mathrm{G}_j(S = s) = \sum_{v_s}^{V_s} n! \prod_{\alpha=1}^{d} \frac{\tilde{p}_{\alpha_j}^{\,r_{\alpha_{v_s}}}}{r_{\alpha_{vs}}!}$$

being the estimation of probability (6), where

$$\tilde{\mathbf{p}}_j = \left(\tilde{p}_{1_j}, \tilde{p}_{2_j}, \ldots, \tilde{p}_{d_j} \right)$$

is the vector of estimations for parameters **p**, defined as

$$\tilde{\mathbf{p}}_j = \frac{\mathbf{z}_j}{nk}.$$

It is almost certain that for $n \to \infty$

$$\prod_{i=1}^{n} \left| c(S = s_i) - P_j^*(s_i, \mathbf{z}_j) \right| < \prod_{i=1}^{n} \left| c(S = s_i) - G_j(S = s_i) \right|.$$

It is possible to use modified likelihood estimates that use numerical methods; for example, to solve the system of equations by the iterative method (Zhao, Xu, Wang, & Lin, 2018), or directly maximize the likelihood function of type (Neal & Hinton, 1998).

The application of numerical methods generates numerous problems because: the convergence of iterative methods requires justification (Aster, Thurber, & Borchers, 2005); the likelihood function can have several local maxima (Fan, Xue, & Zou, 2014); the choice of the moment of terminating the calculations in connection to the required accuracy requires justification (Inglis & Gilmore, 2014); finally, Mengersen, Pudlo, and Robert (2012) and Furber (2016) have noted that the accuracy of computation depends on the sample size.

As such, if any element of the sampling of the given distribution model has more than one decomposition, then in the process of finding estimates of maximum likelihood we face a number of computational problems that call into question the practicality of using estimates of maximum likelihood.

Thus, the unbiased estimates of the probability of distribution (2) by formula (8), determined from observation of the most appropriate estimate, are best.

Similarly, it follows that unbiased estimates (3) of reliability function (9), determined from observation of the most appropriate estimate, are best.

Iskakova and Zhaxybayeva (2018) noted that for distribution (7) a maximum likelihood estimation exists if all the elements of the observations or realizations of a sample have no more than one part in the partition. As is known in practice, some elements in the implementation of the sample has more than one partition and the method for determining the estimation of likelihood is not applicable to this model.

Of course, it is possible to use modified likelihood estimates generated through numerical methods; for example, to solve the system of equations by the iterative method presented in Chichagov (2013), or directly maximize the likelihood function of this type.

The application of such numerical methods generates numerous problems. These include: the convergence of iterative methods requires justification, as noted by Ziskind and Wax (1988); the likelihood function can have several local maxima (Zhao, Xu, Wang, & Lin, 2018); the choice of the moment of terminating the calculations in connection to the required accuracy requires justification (Neal & Hinton, 1998); and finally the accuracy of the computation depends to the sample size.

As such, if any element of the sampling of the given distribution model has more than one decomposition, then, when finding estimates of the maximum likelihood, we run into a number of computational problems that call into question the practicality of using estimates of maximum likelihood.

Thus, unbiased estimates of the probability of distribution (3) by formula (8), determined from observation of the most appropriate estimate, are best.

Similarly, it follows that the obtained unbiased estimates (4) of reliability function (10), determined from observation of the most appropriate estimate, are best.

Description of Practical Application

If we suppose that we need to determine an unbiased estimate of the probabilities of distortions of AgroSmart statistical data with values for August 21 and August 30, 2018 in the Egindykol district of the Almola region of Kazakhstan.

We have the realization of sample (4)

$$\mathbf{y}_1 = (0 \quad 0 \quad 0 \quad 0),$$
$$\mathbf{y}_2 = (1 \quad 0 \quad 1 \quad 0),$$
$$\mathbf{y}_3 = (0 \quad 0 \quad 0 \quad 1),$$
$$\mathbf{y}_4 = (0 \quad 0 \quad 0 \quad 0),$$
$$\mathbf{y}_5 = (0 \quad 0 \quad 1 \quad 0),$$
$$\mathbf{y}_6 = (0 \quad 0 \quad 0 \quad 0),$$
$$\mathbf{y}_7 = (0 \quad 0 \quad 0 \quad 0),$$
$$\mathbf{y}_8 = (0 \quad 0 \quad 0 \quad 0),$$
$$\mathbf{y}_9 = (0 \quad 1 \quad 1 \quad 0),$$
$$\mathbf{y}_{10} = (0 \quad 0 \quad 0 \quad 0).$$

Note that the statistics in (5) are

$$s_1 = 0, \quad s_2 = 2, \quad s_3 = 1, \quad s_4 = 0, \quad s_5 = 1,$$
$$s_6 = 0, \quad s_7 = 0, \quad s_8 = 0, \quad s_9 = 2, \quad s_{10} = 0,$$

and statistic (6) is

$$\mathbf{z} = \sum_{k=1}^{10} \mathbf{y}_k = (1 \quad 1 \quad 3 \quad 1)$$

thus unbiased estimations with uniformly minimal risk for probabilities (1) and (2) are correspondingly defined by (8) and (9) as

$$P^*(\mathbf{X} = \mathbf{x}) = \frac{1}{10^4} \prod_{i=1}^{4} \binom{z_i}{x_i} \binom{n - z_i}{1 - x_i}$$

and

$$P^*(S = s) = \frac{1}{10^4} \sum_{s = \sum_{k=1}^{4} x_k} \prod_{i=1}^{4} \binom{z_i}{x_i} \binom{n - z_i}{1 - x_i}.$$

The unbiased estimator with uniformly minimal risk for reliability function (9) is

$$R^*(S=s)=1-\frac{3!}{10^4}\sum_{0\le t<s}\sum_{t=\sum_{k=1}^{m}x_k}\frac{1}{(3-x_3)!}.$$

The data in Table 1 were obtained using the method for constructing unbiased estimation of probabilities (1) as the formulas in (8).

Table 1. Unbiased estimations $P^*(\mathbf{x}=\mathbf{X})$

x	$P^*(\mathbf{x}=\mathbf{X})$
x=(0; 0; 0; 0)	0.5103
x=(0; 0; 0; 1)	0.0567
x=(0; 0; 1; 0)	0.2157
x=(0; 1; 0; 0)	0.0567
x=(1; 0; 0; 0)	0.0567
x=(0; 0; 1; 1)	0.0243
x=(0; 1; 0; 1)	0.0063
x=(1; 0; 0; 1)	0.0063
x=(0; 1; 1; 0)	0.0243
x=(1; 0; 1; 0)	0.0243
x=(1; 1; 0; 0)	0.0063
x=(0; 1; 1; 1)	0.0027
x=(1; 1; 0; 1)	0.0007
x=(1; 0; 1; 1)	0,0027
x=(1; 1; 1; 0)	0.0027
x=(1; 1; 1; 1)	0.0003

Thus $P^*(\mathbf{X}=\mathbf{y}_1)=0.5103$, $P^*(\mathbf{X}=\mathbf{y}_2)=0.0027$, $P^*(\mathbf{X}=\mathbf{y}_3)=0.0567$, $P^*(\mathbf{X}=\mathbf{y}_4)=0.5103$, $P^*(\mathbf{X}=\mathbf{y}_5)=0.2157$, $P^*(\mathbf{X}=\mathbf{y}_6)=0.5103$, $P^*(\mathbf{X}=\mathbf{y}_7)=0.5103$, $P^*(\mathbf{X}=\mathbf{y}_8)=0.5103$, $P^*(\mathbf{X}=\mathbf{y}_9)=0.0243$, $P^*(\mathbf{X}=\mathbf{y}_{10})=0.5103$.

It is obvious that $s = 0$ if $\mathbf{x}$ = (0; 0; 0; 0);

$s = 1$ if $\mathbf{x}$ = (0; 0; 0; 1), $\mathbf{x}$ = (0; 0; 1; 0), $\mathbf{x}$ = (0; 1; 0; 0), $\mathbf{x}$ = (1; 0; 0; 0);

$s = 2$ if $\mathbf{x}$ = (0; 0; 1; 1), $\mathbf{x}$ = (0; 1; 0; 1), $\mathbf{x}$ = (1; 0; 0; 1), $\mathbf{x}$ = (0; 1; 1; 0), $\mathbf{x}$ = (1; 0; 1; 0), $\mathbf{x}$ = (1; 1; 0; 0);

$s = 3$ if $\mathbf{x}$ = (0; 1; 1; 1), $\mathbf{x}$ = (1; 0; 1; 1), $\mathbf{x}$ = (1; 1; 0; 1), $\mathbf{x}$ = (1; 1; 1; 0);

$s = 4$ if $\mathbf{x} = (1; 1; 1; 1)$.

Then, according to formulas (8) and (9), it follows that

$$P^*(S=0)=0.5103,\ \ P^*(S=1)=0.3888,\ \ P^*(S=2)=0.0918,$$
$$P^*(S=3)=0.0088,\ \ P^*(S=4)=0.0003.$$

The graph of unbiased estimations $P^*(S = s)$ for probability $P(S = s)$ is presented in Fig. 1.

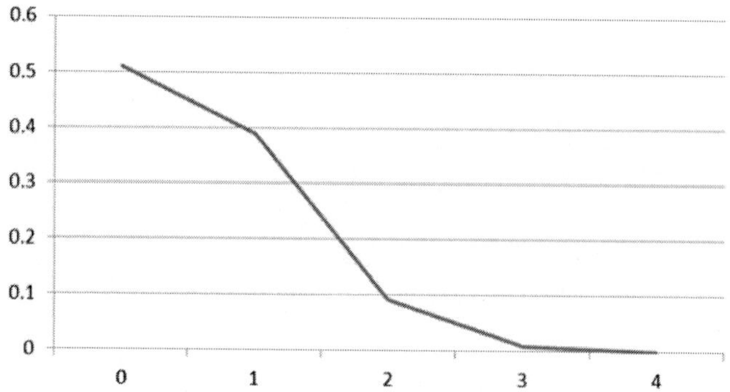

Figure 1. Unbiased estimations for P(S = s)

The data in Table 2 were obtained using a method for constructing unbiased estimation of the probabilities in (1) as presented in the formulas in (3).

Table 2. Unbiased estimations $R^*(\mathbf{x} = \mathbf{X})$

$\mathbf{x}$	$R^*(\mathbf{x} = \mathbf{X})$
$\mathbf{x}$=(0; 0; 0; 0)	1
$\mathbf{x}$=(0; 0; 0; 1)	0.4897
$\mathbf{x}$=(0; 0; 1; 0)	0.4897
$\mathbf{x}$=(0; 1; 0; 0)	0.4897
$\mathbf{x}$=(1; 0; 0; 0)	0.4897
$\mathbf{x}$=(0; 0; 1; 1)	0.2173
$\mathbf{x}$=(0; 1; 0; 1)	0.3763
$\mathbf{x}$=(1; 0; 0; 1)	0.3763
$\mathbf{x}$=(0; 1; 1; 0)	0.2163
$\mathbf{x}$=(1; 0; 1; 0)	0.2163

x=(1; 1; 0; 0)	0.3763
x=(0; 1; 1; 1)	0.0030
x=(1; 1; 0; 1)	0.0010
x=(1; 0; 1; 1)	0,0030
x=(1; 1; 1; 0)	0.0030
x=(1; 1; 1; 1)	0.0003

Unbiased estimators with uniformly minimal risk for reliability function (9) are

$$R^*(S=0)=1,\ \ R^*(S=1)=0.4897,\ \ R^*(S=2)=0.1009,$$
$$R^*(S=3)=0.0091,\ \ R^*(S=4)=0.0003.$$

The graph of unbiased estimations $R^*(S = s)$ for probability $R(S = s)$ is presented in Figure 2.

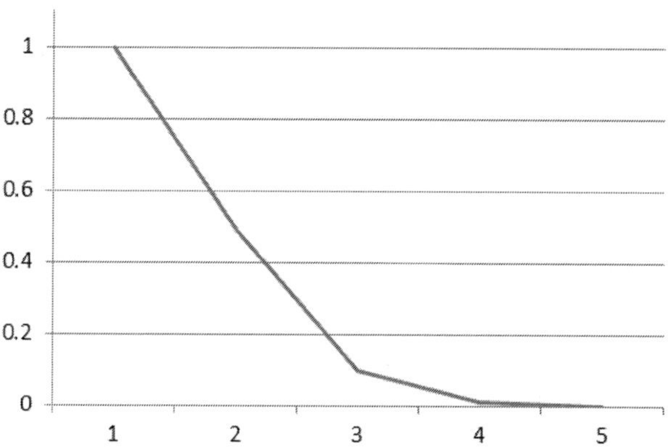

Figure 2. Unbiased estimations for R(S = s)

Thus $R^*(\mathbf{X}=\mathbf{y}_1)=1$, $P^*(\mathbf{X}=\mathbf{y}_2)=0.2163$, $P^*(\mathbf{X}=\mathbf{y}_3)=0.4897$, $R^*(\mathbf{X}=\mathbf{y}_4)=1$, $R^*(\mathbf{X}=\mathbf{y}_5)=0.4897$, $R^*(\mathbf{X}=\mathbf{y}_6)=1$, $R^*(\mathbf{X}=\mathbf{y}_7)=1$, $R^*(\mathbf{X}=\mathbf{y}_8)=1$, $P^*(\mathbf{X}=\mathbf{y}_9)=0.2163$, $R^*(\mathbf{X}=\mathbf{y}_{10})=1$.

Since the all elements of the vector **x** = (0; 0; 0; 0), corresponding to elements of the realization of sample (4), are not less than possible elements of the sum s and the unbiased estimations with good asymptotic properties for reliability are equal to 1. Hence $R^*(y_1)=1,\ R^*(y_4)=1,$

$R^*(y_6)=1,\ R^*(y_7)=1,\ R^*(y_8)=1,\ R^*(y_{10})=1.$

The elements of vector $\mathbf{x}=(1 \quad 0 \quad 1 \quad 0)$ are less than the elements of vectors $(0 \quad 0 \quad 0 \quad 0)$, $(0 \quad 0 \quad 1 \quad 0)$, $(1 \quad 0 \quad 0 \quad 0)$ and an unbiased estimation with good asymptotic properties for reliability is $R^*(y_2)=0.2163$.

The elements of vector $\mathbf{x}=(0 \quad 0 \quad 0 \quad 1)$ and $\mathbf{x}=(0 \quad 0 \quad 1 \quad 0)$ are less than the elements of vector $(0 \quad 0 \quad 0 \quad 0)$. Unbiased estimations with good asymptotic properties for reliability are $R^*(y_3)=0.4897$ and $R^*(y_5)=0.4897$.

The elements of vector $\mathbf{x}=(0 \quad 1 \quad 1 \quad 0)$ are less than the elements of vectors $(0 \quad 0 \quad 0 \quad 0)$, $(0 \quad 0 \quad 1 \quad 0)$, $(0 \quad 1 \quad 0 \quad 0)$. An unbiased estimation with good asymptotic properties for reliability is $R^*(y_9)=0.2163$.

As such, the realization of estimates of the reliability function of the distortions through analysis of a sample are means of intervals ranging from 21.63 % to 100 %. As the value of each realized element from a sample is less than 3, it follows that

$R^*(s_1)=1,\quad R^*(s_2)=0.1009,\quad R^*(s_3)=0.4897,\quad R^*(s_4)=1,\quad R^*(s_5)=0.4897,\quad R^*(s_6)=1,\quad R^*(s_7)=1,\quad R^*(s_8)=1,\quad R^*(s_9)=0.1009,\quad R^*(s_{10})=1.$

Hence, estimates of the reliability of the function of distortions are means of the sum of distortions at intervals ranging from 10.09 % to 100 %. As such, the data from AgroSmart products suggests 100 % prediction, as is often required by customers and other interested parties.

Conclusion

This chapter has presented an analysis showing that data distortions in the AgroSmart system are not particularly high and with an interval of the ratability function estimated as ranging from 27.9 % to 100 %. The distortions of AgroSmart products analyzed used statistical data from

August 21 and August 30, 2018 in the Egindykol district of the Almola region of Kazakhstan; this data was for the off-season in an area with a sharply continental climate. Other data can be processed with the technique described here, which offers an innovative means of analyzing such data.

Acknowledgment

This work was supported by Project AP05135819 of the Ministry of Education, Republic of Kazakhstan.

Bibliography

Iskakova, Aiman, Jumabayeva, Jamilya, Burgumbayeva, Saule, and Bakirbekova, Aigul. 2019. "The Probabilistic Model of Distortions of AgroSmart Data." *Proceedings of the 24th Conference of Open Innovations Association FRUCT*: 632-7.

Khadys, Batyrbyek, Sikhimbayeva, Dinara, and Bozhkarauly, Altai. 2018. "State Regulation of the Development of the Agro-Industrial Complex of the Republic of Kazakhstan." *Journal of Advanced Research in Law and Economics* 9.1 (31): 127-12.

Rustemov, Dastan, Abikayeva, Madina, Rakhimova, G., Omarkozhayeva, and Temirova, A. 2018. "Determining the Efficiency and the Level of Innovative Development in Agriculture: The Case of Kazakhstan." *European Research Studies Journal*, 21(2): 650-15.

Zhou, Fu-li, Wang, Xu, Lin, Yun, He, Yan-dong, and Wu, Nan. 2016. "Influence research of multi-dimensional tech-innovation behavior on tech-innovation performance." *International Journal of Innovation Science* 8.2: 148-13.

Ayman, Iskakova. 2014. "Construction of the most suitable unbiased estimate distortions of radiation processes from remote sensing data." *Journal of Physics: Conference Series*. IOP Publishing, 490 (1): 012113-8.

Ayman, Iskakova. 2016. "Statistical Research for Probabilistic Model of Distortions of Remote Sensing." *Journal of Physics: Conference Series*,IOP Publishing, 738(1): 012004-8.

An, Joonghyun, Seok, Moon Gi, and Park, Daejin. 2016. "Automatic on-chip backup clock changer for protecting abnormal MCU operations in unsafe clock frequency." *IEICE Electronics Express*: 20160808-13.

Jacquesab, Lionel, Béchetc, Eric, and Kerschena, Gaëtan. 2017. "Finite element model reduction for space thermal analysis." *Finite Elements in Analysis and Design* 127: 6-10.

Messerschmitt, David G. 2012. "Interstellar communication: The case for spread spectrum." *Acta Astronautica* 81.1: 227-12.

Ziskind, Ilan, and Wax, Mati. 1988. "Maximum likelihood localization of multiple sources by alternating projection." *IEEE Transactions on Acoustics, Speech, and Signal Processing* 36(10): 1553-8.

Inglis, Matthew, and Gilmore, Camilla. 2014. "Indexing the approximate number system." *Acta Psychologica* 145: 147-9.

Zhao, Bo, Xu, Zhicheng, Xu, Chen, Wang, Caisheng, and Lin, Feng. 2018. "Network Partition-Based Zonal Voltage Control for Distribution Networks With Distributed PV Systems" *IEEE Transactions on Smart Grid* 9(5): 4087-12.

Neal, Radford, and Hinton, Geoffrey E. 1998. "A view of the EM algorithm that justies incremental, sparse, and other variants." *Learning in graphical models*, Springer Netherlands: 355-14.

Aster, Richard C., Thurber, Clifford H., and Borchers, Brian. 2005. "Parameter estimation and inverse problems." *Elsevier Academic.*

Fan, Jianqing, Xue, Lingzhou, and Zou, Hui. 2014. "Strong oracle optimality of folded concave penalized estimation." *Annals of statistics* 42(3): 819-31.

Mengersen, Kerrie L., Pudlo, Pierre, and Robert, Christian P. 2012. "Bayesian computation via empirical likelihood" *Proceedings of the National Academy of Sciences* 110, (4): 1321- 6.

Furber, Steve. 2016. "Large-scale neuromorphic computing systems." *Journal of neural engineering* 13(5): 051001-14.

Iskakova, Aiman, and Zhaxybayeva, Gauhar. 2018. "Maximum likelihood estimates of some probability model of discrete distributions." *Bulletin of the Karaganda University, Mathematics Series* 1(89): 61-9.

Chichagov, V. V. 2013. "On Asymptotic Behavior of Unbiased Probability Estimators for Lattice Distributions with the Mean as a Sufficient Statistic." *Journal of Mathematical Sciences*, 189.6: 932-8.

CHAPTER 2.3

CASE STUDY OF MOBILE VIDEO SERVICES

NIKITA BAZHENOV, ARTUR HARKOVCHUK,
NIKITA BESEDNIY, VSEVOLOD AVERKOV, DMITRY KORZUN
Petrozavodsk State University
bazhenov@cs.petrsu.ru, harkovch@cs.petrsu.ru, besednyi@cs.petrsu.ru,
averkov@cs.petrsu.ru, dkorzun@cs.karelia.ru

Abstract. Video data provides a very rich set of information for use in knowledge-based systems. In this chapter, we summarize pilot demonstrations of video vision services that we have developed. This summary describes an innovative knowledge-based system, where an essential part of the knowledge is derived from video data. These data were captured online using everyday cameras: embedded, web, or mobile. We suggest that this case study shows a particular way of designing edge-centric computing systems with a potentially unlimited spectrum of smart services for the end-users.

Introduction

Using various edge devices, a video vision system can utilize many video data sources—online cameras—which are embedded in the environment, attached as web cameras to the Internet, or even mobile (on a smartphone). Each video flow can be data mined based on existing methods of video-analytics and the implemented algorithms (e.g., see Montori et al. 2018 and Korzun & Balandin 2016 for example services and background methods).

In this chapter, we summarize four demo video services created by us and presented at the FRUCT24 Conference in April 2019 (see the program of the demo session at fruct.org/program24). The demo services cover various application domains of video-analytics.

1. Indoor Service for Human Recognition (Bazhenov et al. 2019).
2. Semantic Search Service by a Person's Facial Photo (Harkovchuk & Korzun 2019).

3. Interaction Service Using On-screen Visual Information (Besedniy & Korzun 2019).
4. Recognition of Human Interaction with Objects (Averkov & Kulakov 2019).

The innovative video vision system was created with the goal of the four developed services becoming seen as base services. They can be used in combination to construct advanced services for end-users within the same environment. This service composition approach can be implemented using smart space technology. In particular, Vdovenko et al. (2015) have shown how an existing set of services can be extended through service composition.

The rest of this chapter is organized as follows. First, we consider related work in the area of smart systems with video-based services. Then we introduce our concept of a smart video vision system integrating basic services. Each subsequent section is devoted to the designated base services derived from the four described. In conclusion, we provide a summary of this study.

Related Work

A smart video vision system can be created as a smart space (Korzun et al. 2014). The constructed services are based on information sourced from various devices, including cameras. Our reference example is the SmartRoom system where a group of people collaborate in a virtual shared workspace using the provided mobile services to accelerate information exchange, interpretation, and subsequent knowledge generation.

Service composition and enhancement is discussed in Vdovenko et al. (2015). The users of the SmartRoom system can visually develop their own social programs when working together.

Such a collaboration using the Internet of things (IoT) environment is discussed in Montroi et al. (2018). The key problem at hand is how to combine the data from many devices (i.e., from persistent monitoring of the environment) and find appropriate services to process this data.

A key-frame architecture based on video surveillance is discussed in Gao et al. (2018). This paper considered a method for detecting a key-frame in video using f-divergence techniques. The detection of essential events in video flows occurs in real time, and surveillance summarization is constructed as a service.

A comprehensive survey of video interaction tools is presented in Schoeffmann et al. (2015). This study considers how to manipulate direct video data and visualize video content using low-cost devices that are mass-produced and readily available on the market.

This chapter provides a case study based on several video services demonstrated at the FRUCT24 conference in 2019. Let us consider the services in respect of similar work in the area of video services.

Indoor Service for Human Recognition (Bazhenov et al. 2019). A similar service for device-free human presence detection using frequency domains is discussed in Mrazovac et al. (2015).

Semantic Search Service by a Person's Facial Photo (Harkovchuk & Korzun 2019). A similar technique for semantic search is discussed in Khan and Malik (2018).

Interaction Service Using On-screen Visual Information (Besedniy & Korzun 2019). A similar problem of human-to-human interaction is discussed in Liu et al. (2017), based on spatial and motion trends.

Recognition of Human Interaction with Objects (Averkov & Kulakov 2019). An appropriate machine learning technique is discussed in Chao et al. (2018) for the detection of human-object interactions.

Smart Video Vision System

Our case study considers a generic smart video vision system that combines the following service classes.

Presence detection based on facial recognition. Such a service provides basic information on people within a specified area. Many key individual characteristics of a person can be identified (e.g., full name, unique facial profile).

Information search extended to open Internet resources. Such a service provides additional information about particular people (using their photo and video data). In particular, many context details can be discovered (e.g., a person's interests, Facebook profile).

Human interaction tool using multimedia screens. Such a service supports human-to-human interaction by displaying visualized information for human decision-making. In particular, screens can be public (e.g., wall-mounted board, TV display) or personal (e.g., smartphone, tablet).

Activity recognition based on human movement detection. Such a service implements video data mining for analysis of human-to-human and human-to-object interaction. In particular, various complex events can be detected and serve as an initial event to facilitate the construction of further services as a reaction to the event.

Our concept of a generic smart video system is shown in Fig. 1. This generic class of systems supports advanced service construction based on the four base classes considered. A promising reference system case is the SmartRoom system (Korzun et al. 2014), where the room is equipped with multiple real-time cameras connected to the system.

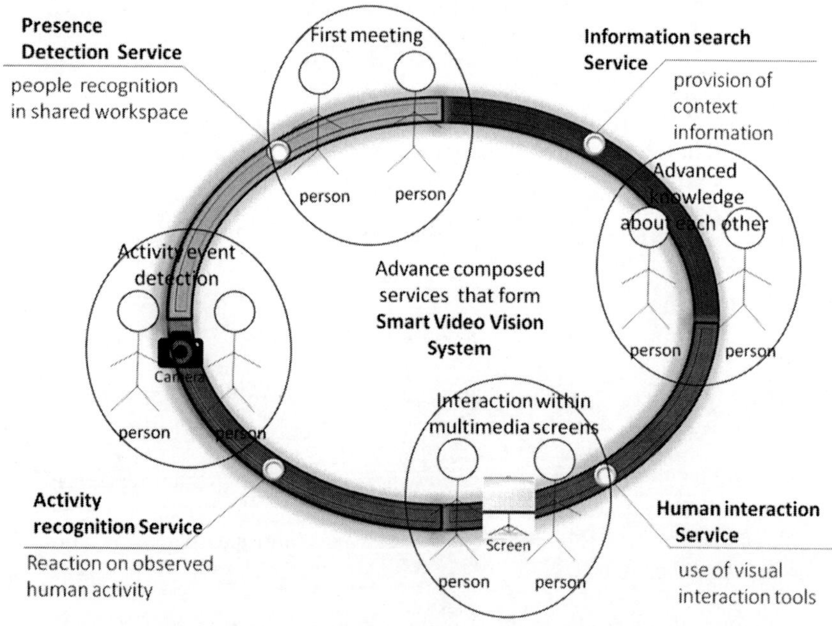

Figure 1. Smart video vision system composed of base services

A presence detection service identifies human participants and allows them to interact within the information space. The information search service provides additional information from open Internet resources about human participants using basic identification information derived from the presence detection service. The human interaction service enables visually driven interaction using available multimedia equipment. The activity recognition service uses data from several online cameras to analyze the ongoing activity in the room, i.e., detecting starting points for proactive construction of appropriate services to solve situational end-user needs.

People can be divided into several groups and each group performs its own activity. Such activities can display an informational overlap if needed. The basic participation tools are personal mobile devices, such as smartphones and tablets. Most of the initial data come from online cameras (i.e., in real-time mode), including smartphone cameras. The services extract information from the data so that a shared information space can be formed for collaborative work. The computational load is on one or more edge servers. Part of the computation can be delegated to appropriate local devices, e.g., smartphones (thus making a smartphone camera "smart").

The division into several services (or a service with microservices) offers the promise of a multi-service architecture (Nagothu 2018). This kind of architecture increases accessibility and durability and can potentially make use of blockchain technology.

The rest of this chapter describes each proposed service class. We expect that the combined use of these basic services can lead to a wide spectrum of advanced video-based services for end-users.

Presence Detection Service

Human presence detection is typically based on human recognition. This requires expensive computational resources and digital equipment. Nevertheless, our demo (Bazhenov et al. 2019) has shown that a low-cost indoor service can be implemented using simple everyday cameras and small, low-power computers (e.g., Raspberry Pi or smartphone).

The service keeps a registration list of people (persons present in the room). The camera captures a person's face as it appears in the room. Every detected face, whether the person is registered (recognized) or not, is checked.

The system shares and maintains information about all human participants present in the smart space. This information can be used to construct various services for end-users, including personalized services (Korzun & Balandin 2016).

Initial information for the service is provided by user profiles. Each person registers in advance on the system, using an appropriate login&password mechanism. An advanced user profile can include such parameters as name, photo, gender, age, interests, and registration on social networks.

The service can be implemented using a small onboard computer that communicates with a cheap everyday camera (wired or remote). In particular, analysis of how many people are in an area is easily implemented using existing face recognition algorithms. An example scheme employing a Raspberry Pi device is shown in Fig. 2.

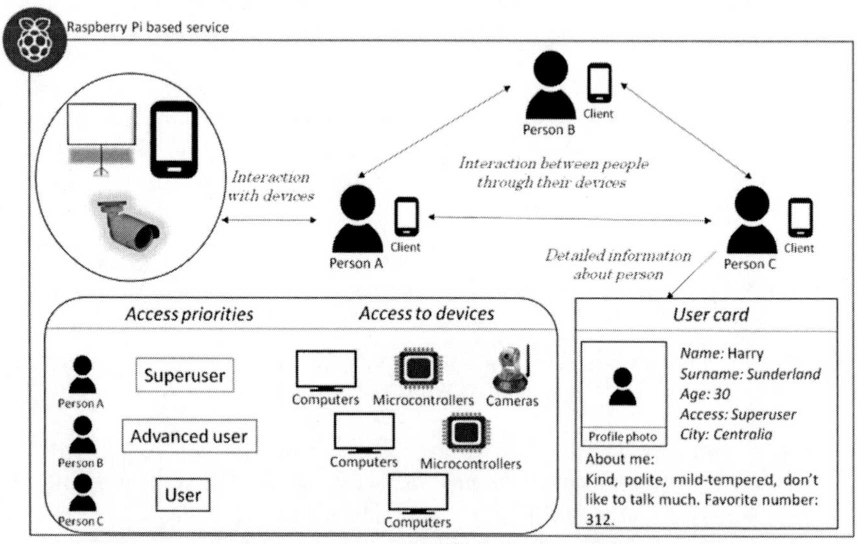

Figure 2. Scheme for presence detection using a small onboard computer

The service can support such functions as:

- Detection of several people in the same area, so that they can interact;
- Differentiation of participants who are present according to access priorities;

- Access control to various resources in physical space (e.g., floors, computers, door-locks, printers);
- Provision of detailed information about participants (e.g., name, photo, interests).

Information Search Service

This service searches for information about a person in open Internet sources. It requires significant computational resources and is usually unavailable to common users of computing devices. Nevertheless, our demo (Harkovchuk & Korzun 2019) showed that some information searches can be implemented using low-power end-user devices. In particular, one can use a facial photo to find a lot of information out about a person.

Let us assume that the initial data for the search request is a person's facial photo. In particular, it can easily be made from a smartphone or using a presence detection service (see the previous section). Other information from the user profile can be used to augment the photo.

Social networks can be used to find a lot of information about a person since users often add photos and descriptions of themselves. Many existing search engines, social networks, and databases can be used for such a search. The problem remains of how to analyze the data to develop knowledge about a particular person in the system. The proposed scheme is shown in Fig. 3.

Various edge servers can be used to perform expensive computations and store data. Searches are iterative, i.e., found information is input for use in the next search round. As a result, many web pages are analyzed to find details associated with a given person.

Moreover, this search technique supports ranking and found units of information can be characterized with a numerical value. This can quantitatively reflect the relevance of the found information. For instance, the more times a detail is found during an iterative search, the more relevant it is.

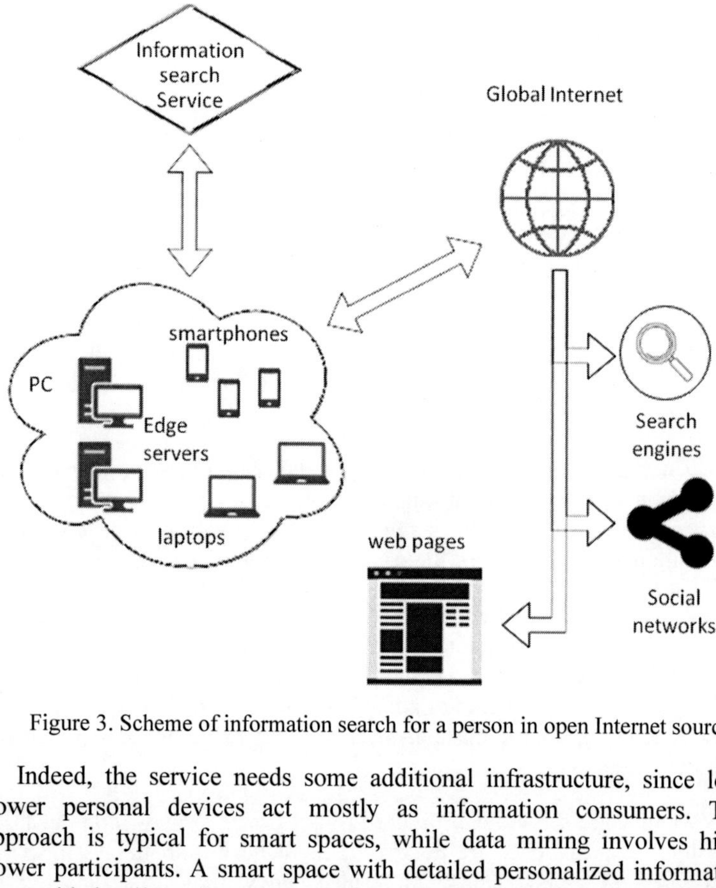

Figure 3. Scheme of information search for a person in open Internet sources

Indeed, the service needs some additional infrastructure, since low-power personal devices act mostly as information consumers. This approach is typical for smart spaces, while data mining involves high-power participants. A smart space with detailed personalized information is provided and service clients deliver relevant information to end-users for further analysis and decision-making (Korzun & Balandin 2016).

Human Interaction Service

This service assists with collaborative work in a shared information space (in other words, organizing a smart space or workspace). The problem relates to the organization of effective information exchange between participants when one participant has knowledge that can be used by another to produce new knowledge. Our demo (Besedniy & Korzun 2019) offered a solution to the particular case of this generic problem: the speaker can extend her/his slides during the presentation to provide more focused information to other participants.

Various multimedia equipment is used to visually share information. In particular, wide-screens, interactive boards, projectors, and user mobile devices (such as smartphones, tablets) can be used to facilitate interaction between human participants. The latter operate with displayed objects. As such, the service shortens the distance between information (or knowledge) exchanged between participants.

The scheme is shown in Fig. 4. Participants can use free space on any screen in the room. They can form a group for brainstorming using a wide-screen or collaborative work using a virtual screen. This implementation of the blackboard model supports users in dynamically changing or supplementing displayed information. In particular, users can accompany their questions with information during presentations. During discussion, this information increases the efficiency of the discussion and improves the answers provided.

To give an example, Human group A uses a screen connected to a PC/laptop. On this screen, members can edit (add, modify, remove) the displayed information. For Human group B, the service provides a common virtual screen. Their devices use the shared information space to synchronize their screens. All users of the group can add media information to their screen. Other group members can observe this information after the device is synchronized with other devices in the group.

The service can be used to organize collaborative work within the SmartRoom system (Korzun et al. 2014). This system supports events such as lectures, seminars, and conferences. One interesting service case involves a feedback service for lectures. For example, a lecturer can give their presentation using slides while conducting a survey to understand how well other participants are receiving the information. Using a smartphone the participant can answer some basic questions on the material being studied.

The feedback is analyzed to estimate the level of understanding and this estimate is placed online so the lecturer can analyze it. For instance, if the estimate shows a low level of understanding, the lecturer needs to ensure greater focus on the material presented.

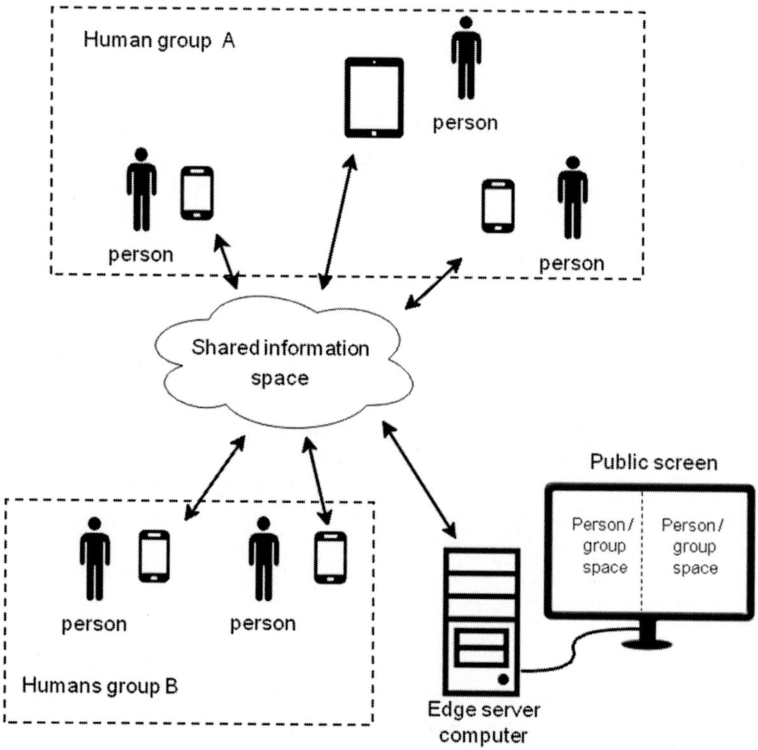

Figure 4. Human interaction based on visual information sharing on public and personal screens

Activity Recognition Service

This service provides a rich source of contextual information about human physical activity. People make movements, reflecting their interaction with other persons or equipment (surrounding physical objects). This type of activity can be monitored in the working area using embedded or mobile cameras. Our demo (Averkov & Kulakov 2019) showed that the following facts (basic events) can be effectively detected:

- a person appearing in the area;
- a person who is close to a given object.

Recognition of people can also be made using of neural network methods, as shown in Fig. 5. Data analysis of video images from a given camera can be undertaken using a trained neural network. In particular, we can use the Caffe tool for neural network construction, training, and analysis. OpenCV and Python can also be used for standard operations with video data.

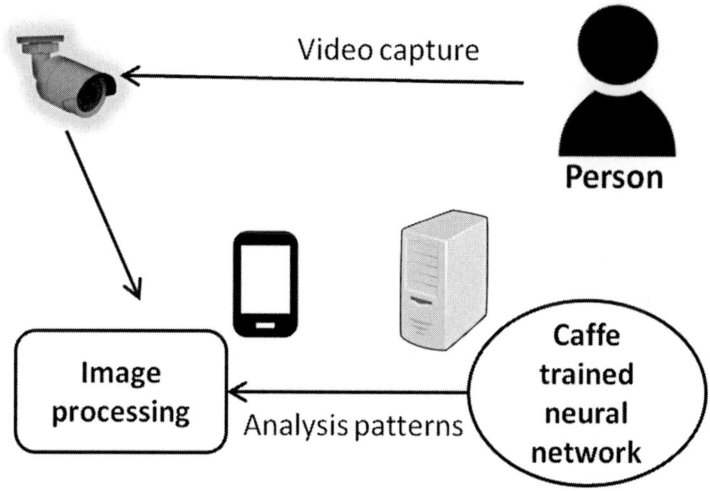

Figure 5. Scheme of human activity recognition based on video flow from camera

The key problem relates to quick recognition of active events and then their composition into complex events. Detection of the above two basic events enables recognition of such short-term events as

Person A being near object B from time T_1 until time T_2.

Furthermore, the service can evaluate the behavior of A while near B in $[T_1, T_2]$. A distance metric $p(t)$ is associated with the event to show the closeness of A to B in time t.

Conclusion

This chapter has summarized four demo video-data services presented at the FRUCT24 Conference in April 2019. We have shown that these services can be integrated to develop an innovative knowledge-based system. The system enables the construction and delivery of many new smart services to end-users. Importantly, the four basic services provide context information about people and their interactions, which is shared among all system participants, thus enabling the use of various edge devices to perform data processing and service construction.

Acknowledgment

This research was supported by the Russian Fund for Basic Research under research project # 19-07-01027. The results were implemented by the Government Program of Flagship University Development for Petrozavodsk State University 2017–2021.

References

Averkov V. and K. Kulakov, "Application for determination of user interaction with the objects using a camera," in Proc. 24th Conf. Open Innovations Association FRUCT, Apr. 2019, p. 815.

Bazhenov N., A. Harkovchuk, S. Marchenkov, and D. Korzun, "A low-cost indoor service for human recognition," in Proc. 24th Conf. Open Innovations Association FRUCT, Apr. 2019, pp. 816–817.

Besedniy N. and D. Korzun, "An interactive presentation service for smartroom," in Proc. 24th Conf. Open Innovations Association FRUCT, Apr. 2019, pp. 818–820.

Chao Y., Y. Liu, X. Liu, H. Zeng, and J. Deng, "Learning to detect human-object interactions," in 2018 IEEE Winter Conference on Applications of Computer Vision (WACV), March 2018, pp. 381–389.

Harkovchuk A. and D. Korzun, "Semantic information search service by person's face photo," in Proc. 24th Conf. Open Innovations Association FRUCT, Apr. 2019, pp. 821–823.

Gao Z., G. Lu, C. Lyu, and P. Yan, "Key-frame selection for automatic summarization of surveillance videos: a method of multiple change-point detection," Machine Vision and Applications, vol. 29, pp. 1101–1117, Oct 2018.

Khan A.A. and S.K. Malik, "Semantic search revisited," in 2018 8th International Conference on Cloud Computing, Data Science Engineering (Confluence), Jan 2018, pp. 14–15.

Korzun D. and S. Balandin, "Personalizing the Internet of Things using mobile information services," in UBICOMM 2016: The Tenth International Conference on Mobile Ubiquitous Computing, Systems, Services and Technologies, 2016, pp. 184–189.

Korzun D., I. Galov, A. Kashevnik, and S. Balandin, "Virtual shared workspace for smart spaces and M3-based case study," in Proc. 15th Conf. of Open Innovations Association FRUCT, Apr 2014, pp. 60–68.

Liu B., H. Cai, X. Ji, and H. Liu, "Human-human interaction recognition based on spatial and motion trend feature," in 2017 IEEE International Conference on Image Processing (ICIP), Sep. 2017, pp. 4547–4551.

Montori F., L. Bedogni, and L. Bononi, "A collaborative Internet of Things architecture for smart cities and environmental monitoring," IEEE Internet of Things Journal, vol. 5, no. 2, pp. 592–605, April 2018.

Mrazovac B., B. M. Todorovi´ c, D. Kukolj, and M. Temerinac, "Device-free human presence detection using frequency domain," in 2015 IEEE 1st International Workshop on Consumer Electronics (CE WS), March 2015, pp. 23–26.

Nagothu D., R. Xu, S. Y. Nikouei, and Y. Chen, "A microservice-enabled architecture for smart surveillance using blockchain technology," in 2018 IEEE International Smart Cities Conference (ISC2), 2018.

Schoeffmann K., M. A. Hudelist, and J. Huber, "Video interaction tools: A survey of recent work," ACM Computing Surveys (CSUR), vol. 48, no. 14, Sep 2015.

Vdovenko A., S. Marchenkov, and D. Korzun, "Enhancing the SmartRoom system with e-tourism services," in Proc. 17th Conf. Open Innovations Framework Program FRUCT, Apr. 2015, pp. 237–246.

CHAPTER 2.4

USING AN OBJECT MODEL IN THE IMPLEMENTATION OF COOPERATIVE SOFTWARE AGENTS ON THE SMART-M3 PLATFORM

KIRILL KULAKOV
Petrozavodsk State University
kulakov@cs.petrsu.ru

Abstract. Creating software systems, using specific software tools, is time-consuming. The paper discusses the development of intelligent systems using the Smart-M3 platform. The platform provides an API for interaction using request-response and subscription-notification mechanisms based on triples. At the same time, software agents use object models for their work. This paper presents an approach to the organization of a layer between a software agent and an API using the generation of source code in the Java programming language. Problems of data consistency and the implementation of a low-loaded notification model are considered.

Introduction

One of the foremost areas of information technology is the development of smart services. Smart services combine the capabilities of modern algorithms with those of software and hardware architectures. However, the creation of smart services requires the use of specialized software tools, leading to an increase in the complexity of implementation.

The Smart-M3 platform is a tool for building smart services (Korzun 2011). Software and hardware solutions based on the Smart-M3 platform have a distributed architecture and consist of knowledge processors (KP) and one or more semantic information brokers (SIB). An SIB stores information in the form of a set of triples: "object-predicate-subject". The interaction between KPs occurs through the SIB: one KP publishes information and a second KP receives it (Honkola 2010). At the same time, relations between KPs can be quite complex; for example, one group of KPs waits for data to come from a set of another group of KPs. As such, the actual task centers on how to organize the information space of the SIB and methods of KP interaction (Galov 2014).

A popular way of presenting the structure of the SIB information space is as an ontological model (Lomov 2013). This model allows us to describe the associated data elements in the form of objects and build the required relationships between them, data properties, and instances of objects, etc. (Davies 2006). There are some generally accepted models for a number of areas and a service ontology model may include parts of third-party models. For example, location data can be represented using Basic Geo (WGS84 lat/long) Vocabulary (Brickley 2004).

For the organization of communication between KPs and SIBs, a large number of solutions exists. The most interesting one uses a code generation mechanism to provide access to high-level functions. In this case, the library includes some additional functions; for example, reconciliation with the dictionary or representation data in the objects or structures (Korzun 2011).

We propose a SmartJavaLog project (Kulakov 2019) for generating code based on Java and using an ontological model. Low-level tasks, such as converting triples to and from an object, sending change notifications, and tracking changes in object instances, are included in the generated object code and executed automatically. Thus, KP implementation is performed in a traditional object model using the capabilities of Java. This extends previous work presented in Kulakov (2019).

A key problem in implementing KPs is how best to organize interaction between a KP and a SIB. Often, a KP does not use the whole information object, but only a part of it. Thus, it is necessary to provide tracking of changes in the required data. A particular example is the task of analyzing the properties of an object after changing one of the "sign" fields. As a rule, this problem is solved by placing an additional load on the SIB: each time a "sign" is received, the current instance of the object is loaded from the SIB using either one or several requests (Vdovenko 2014).

Another problem is that KPs can use the same object in different parts of the code. This problem is especially acute when asynchronous programming is used and the object is required in different callback functions. This may lead to an object being "out of sync" with the current version in the SIB, requiring the organization of subscriptions for each instance, increasing the load on the SIB (Vdovenko 2017).

One solution to both problems is to organize a centralized local repository for each KP. In this case, only one subscription is required between the repository and the SIB; references to the repository objects are used inside the KP. Using code generation based on the ontological model in conjunction with a centralized local repository allows a reduction in the work required for KP development.

The rest of this chapter is organized as follows. "Related Work" contains a summary of current developments in this area. "KP Interaction Model" describes the interaction methods and smart notifications used. "Centralized Object Storage Model" presents the storage model approach and common scenarios. "Implementation" describes a proposed source code generator with code samples. Finally, "Conclusion" summarizes the chapter.

Related Work

The implementation of the interaction between a KP and SIB can be carried out in two ways. This can be through low-level programming: KPI_low, Java-KPI, and python KPI (see Table 1); or high-level programming: SmartSlog (Lomov 2013), Smart-M3 ontology to C-API generator, and Smart-M3 ontology to the Python generator (see Table 1). As a rule, high-level programming is based on one of the low-level libraries.

The library for low-level programming implements the SSAP protocol and allows access to the basic functions of the SIB, including: adding or removing triples; sending queries; and subscribing to notifications. Control over the correct representation of data in the form of triples, the correspondence of the ontological model, and the determination of the interaction between KPs rests with the KP developer.

Table 1. Software, tools and other resources

Name	Description	Link (URL)
KPI_low	KPI_low is an interface for implementing Smart-M3 Knowledge Processors in ANSI C	https://sourceforge.net/projects/kpilow/
Java-KPI	Java-KPI is an interface made by the University of Bologna and VTT (Oulu) for interacting with the Semantic Information Broker of Smart-M3	https://sourceforge.net/projects/smartm3-javakpi/
Python KPI	Python KPI is an interface for interacting with the Semantic Information Broker of Smart-M3, made by SOFIA project	https://github.com/iddi/sofia/wiki/Download-smart-m3-python-kpi
Smart-M3 ontology to the Python generator	Smart-M3 ontology for the Python generator is a tool for ontology-based source code generation for the Python language	https://sourceforge.net/projects/smart-m3/
The OWL-DL to C(glib) API generator	OWL-DL is a Web Ontology Language intended to provide a language that can be used to describe the classes and relations between them. The goal for the generator is to generate a C API, abstracting the ontology and SIB communication	https://sourceforge.net/projects/smart-m3/
SmartJavaLog	SmatJavaLog is ontology-based java code generator for Smart-M3 platform	https://github.com/seekerk/smartjavalog
OWL API	OWL API is a Java API for creating, manipulating and serializing OWL Ontologies	https://github.com/owlcs/owlapi

There is a large amount of work on developing computer-aided programming tools to automate and simplify the task of ontology-driven software development. A related popular research area is that of mapping between ontologically-derived concepts and object-oriented language constructs. Evermann and Wand (2005) propose mapping rules to guide the construction of object-oriented conceptual models from ontological conceptual models. This work indicates that object-oriented languages are expressive enough to model real-world application domains. Batanov and Vongdoiwang (2007) consider the similarities and differences between object models and ontologies. The most substantial difference is that while ontologies represent well-structured descriptions of mutually related terms, the object model represents a system as a structure of objects ready for implementation in software.

Another popular research area is that of creating software based on a code generation approach. This approach is commonly used for dynamically typed, interpreted, and object-oriented program languages (e.g. Java). The Protégé source code generator (Knublauch 2004) is a module for a Protégé open-source ontology editor. It generates a factory that can serve as the entry point to the generated code, providing access to existing individuals in the ontology and the ability to create new individuals in the ontology. The Vocabulary class collects all used IRIs. Each object is represented as an interface with list of methods.

The OWL-DL to C(glib) API generator (see Table 1) generates a C API, abstracting the ontology and SIB communication. The KP template can be used by the developer to easily start using the generated API. It contains all the code needed to join the SIB. The code can be inserted into the kpMainLoop. The generated code includes functions and structures for the KP developer to access the generated ontology mappings (generic.h, generic.c, etc.) and a mediator for accessing the SIB that also contains a local triple store (mediator.h, mediator.c).

RDFReactor (Volkel 2005) is an open-source code generator that transforms a given ontology in a RDF Schema into a familiar, dynamic, object-oriented Java API (classes and etc.). At runtime, objects of these classes act as stateless proxies on the RDF model. This enables developers to interact with java proxy objects, thus allowing them to stay in their own world and at the same time make use of the advantages of RDF capabilities.

SmartSlog (smart space ontology) (Korzun 2010) is a software/application development kit (SDK or ADK) for programming Smart-M3 agents (KPs) that consume/produce smart space content according to high-level ontological representation. SmartSlog applies the code generation approach. Given an OWL ontology description, the ontology programming library is produced. This provides an API for accessing the smart space via a Smart-M3 semantic information broker (SIB) and data structures and functions to represent and maintain locally in KP code all ontology classes, relations, properties, and individuals. Despite the use of classes, the interaction with the SIB is carried out using triples.

The problem of managing objects and controlling data integrity is a classic one in object-oriented programming. There are various design patterns (Hannemann 2002), such as a singleton or factory that allows the management of object data. However, these templates require implementation through program code.

There are various software tools and approaches to implementing the interaction of smart service software agents. However, they either do not consider the concept of object-oriented programming, or are not oriented to use with the Smart-M3 platform.

KP Interaction Model

The interaction between KPs is a key function of the smart service. The general approach offers interaction through a semantic information broker (Honkola 2010). The semantic information broker provides an application programming interface for two interaction methods.

- "Query—Answer" interaction. The KP sends a request for the required data and the SIB receives all triples found.
- "Subscription—Notification" interaction. The KP subscribes to a specific dataset and the SIB sends all changes until the KP unsubscribes.

The SIB supports two types of query: a basic query (defining a triple part) or an extended query in the SPARQL language. The basic query has a strict limitation: the subject, the predicate, and the object can be precisely defined or take any value. A SPARQL query should return the result as a set of triples i.e. the query should have three return values.

The SIB provides a subscription using a triple template or a SPARQL query (not supported in some SIB implementations). The subscription operation has the same restriction as the query operation. After the subscription operation is completed, the SIB sends the first notification with all the triples found (after the query operation). When data changes in the SIB, the KP receives a notification. Each notification, except the first, includes a set of new or modified triples and a set of deleted triples.

In practice, the implementation of the subscription operation has additional restrictions. If the response or notification contains a large set of triples (for example, after the query "give me all triples") the SIB can be overloaded (Vdovenko 2017). A knowledge processor should also not use a large number of subscriptions because the SIB can be overloaded. As a result, the traditional “get the data when you need it” approach (see the example in Fig. 1) is not applicable in practice. For example, the KP creates three object instances with property value “2”, then, instance #2 sets property value to “3”, instance #3 sets property value to “5” and sends an update query. If an object instance has a subscription, it will receive a notification with a new property value (“5”). But other instances of the object, without a subscription, do not find out about the changes taking place.

One solution is to use “smart notifications” (Galov 2014) when the KP sends additional “notification” triples to the SIB with object changes. Other KPs can subscribe to receive “notification” triples. If the SIB notifies it of new “notification” triples, then the KP can make a query to obtain all individual data.

There are a number of studies devoted to the problem of constructing smart spaces based on ontological models. “Enhancing the SmartRoom System with e-Tourism Services” (Vdovenko 2015) presents an ontological representation of the SmartRoom system with the e-tourism service information space. “An Approach for Monitoring and Smart Planning of Urban Solid Waste Management Using Smart-M3 Platform” (Catania 2014), presents an ontological model for the smart planning of the urban solid waste management service. “Ontological Model of Multi-source Smart Space Content for Use in Cultural Heritage Trip Planning” (Kulakov 2015) describes the ontological model for the cultural trip planning service. It may be noted that most ontological models are represented in the form of a tree-like structure, using the terms “object” and “object property”.

Despite the presence of different ways of interacting with the SIB, implementation in practice requires non-trivial solutions and the consideration of various restrictions. These features can be independently implemented by the developer KP each time, but it is preferable to use ready-made libraries and code templates with already implemented features.

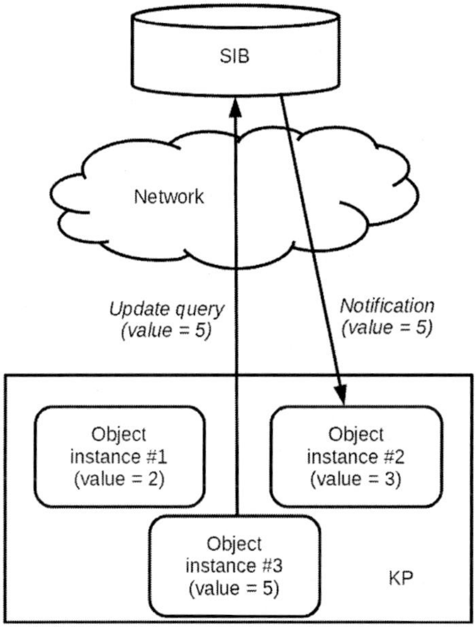

Figure 1. Example of data flows for equal object instances in KP

Centralized Object Storage Model

One of the ways of solving the problem of access restrictions to the SIB is to create a local centralized data storage (see Fig. 2). Centralized storage minimizes interaction with the SIB by reducing duplicate subscriptions and eliminating the need for repeated requests for these objects.

The disadvantages of local centralized storage include the need to control the relevance of the information used and the need to store redundant data. However, the smart subscription mechanism discussed earlier allows monitoring of all changes to the required objects to be organized without additional resource costs. Loading only the required objects and periodically cleaning the local storage will minimize storage redundancy.

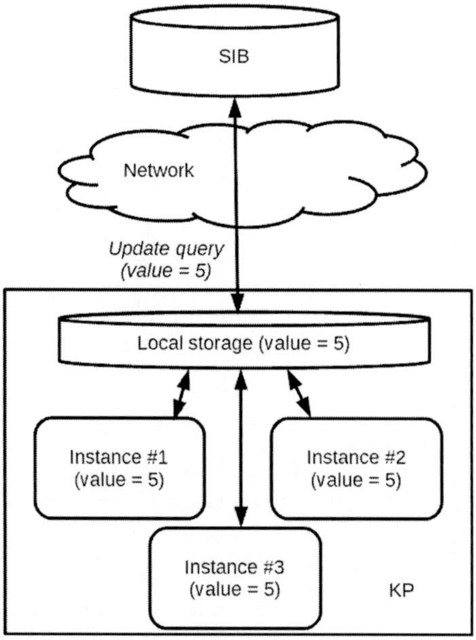

Figure 2. Example of using local storage in KP

In connection with the use of code generation, local storage should ensure the universality of data storage, regardless of the type of object. Thus, it is optimal to use a structure similar to the SIB-data representation in the form of "object-predicate-subject" triples. The URI of the object field uniquely identifies the identity of the triple with the object, the predicate field determines the property of the object, and the subject field contains either a simple property value or the identifier of another object. To identify the type of object, a special triple is used with the value of the predicate "http://www.w3.org/1999/02/22-rdf-syntax-ns#type".

Consider the typical scenarios of working with the local storage.

1. A request for receiving object data. If there are no triples in the local storage with the required identifier, a request is sent to the SIB. The results of the query from the SIB are stored in the local storage and are returned in response.
2. A request to change data. Changes are made to the local storage and the changes are subsequently sent to the SIB.
3. Notification of the appearance of new triples in the SIB. Upon receipt of new triples, local storage is updated. In a case when the

object identifier is not tied to the type (there is no special triple), a request is made to the SIB.

As a result, it is possible to integrate local storage in the form of an intermediate layer between the KP and SIB, ensuring the relevance of local storage data and reducing the load on the SIB. As such, local storage offers a solution to the problem of data consistency.

Implementation

The ontology-based object model is implemented by a source code generator (SmartJavaLog) (see Table 1). The ontological description format is very flexible. Object properties can be represented as a set of axioms. Therefore, we need a two-step algorithm: analysis of data from ontology and generation of source files. The generation process has the following steps (see Fig. 3):

1. parse ontology file;
2. generate files with common Java source code;
3. generate files with ontology specific Java source code.

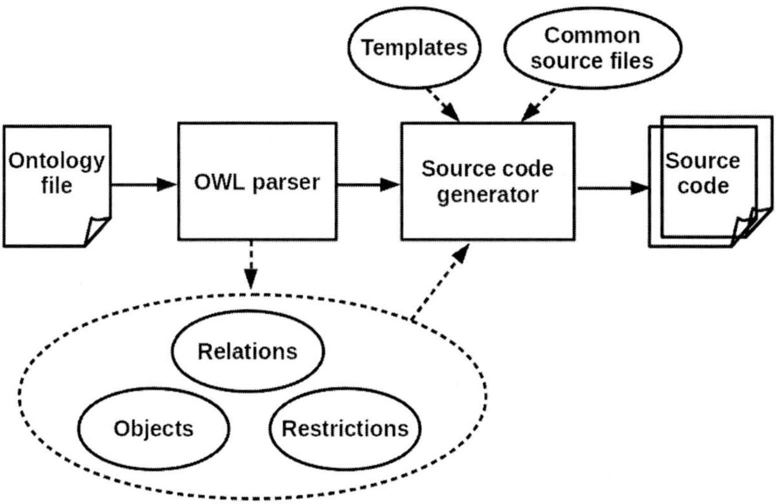

Figure 3. Source code generation process scheme

We use the OWL API library to parse ontology. The library uses "Visitor" pattern (implemented in OntologyVisitor class, see Fig. 4) and presents the ontology file as a set of nodes and axioms. All gathered data is stored in a single instance of OntologyFactory class ("Factory" pattern). OntologyFactory includes Maps of objects (OntologyObject class); properties (OntologyProperty class); types (OntologyComplexDataType class) and comments. The IRI uses a key in map structure.

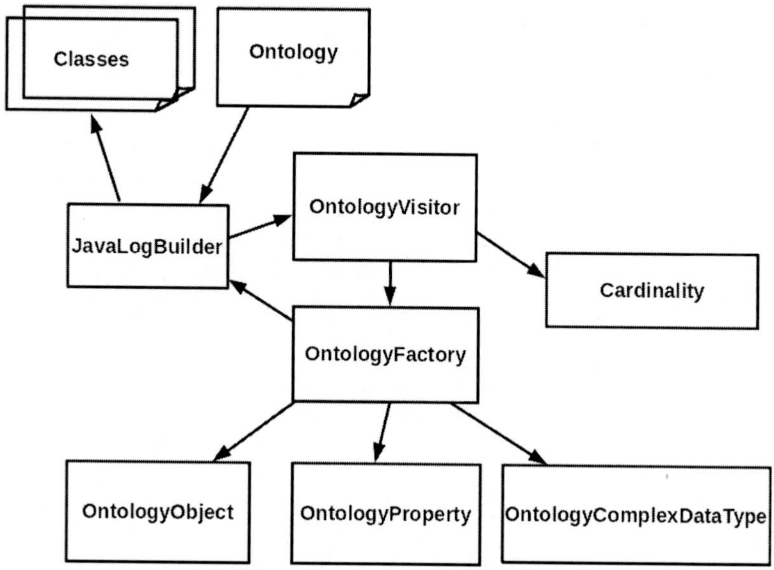

Figure 4. SmartJavaLog high-level architecture

Common java source code uses JavaKPI library for interaction with the SIB and implements core mechanisms, such as asynchronous work, interfaces, and factories. The common java source code includes the following items:

- class BaseRDF—parent class for all ontology objects, includes triples store, work with listener methods, and methods for interaction with the SIB;
- class KPIproxy—JavaKPI library wrapper, implements asynchronous adding and removing of triples, join, leave and query processes;
- interface QueryListener—notification interface, includes add item notification;
- class SIBFactory—main point to work with one or more SIBs, uses "Factory" template;

- class SIBQueryTask—parent class for asynchronous access to the SIB;
- class SIBSubscribeTask—parent class for subscription processes;
- class SubscribeQuery—main point of subscriptions, implements wrapper for JavaKPI subscriptions;
- interface SubscribeListener—subscription notification interface;
- class TaskListener—parent class for all tasks;
- interface UpdateListener—object changes notification interface.

During the implementation of a KP for Android, we were faced with a significant limitation: any work with the network should not be in the main thread. The original version of the Java KPI does not use threads for communication with the SIB. Therefore, SmartJavaLog implements wrappers for the Java KPI function. The wrapper creates a SIBAsyncTask (extends AsyncTask class) based thread and call Java KPI function. When the Java KPI function ends, the SmartJavaLog notifies all listeners. Another problem is that non-android platforms do not have an AsyncTask class. For these platforms, we implement a simple similar class with single-thread work.

Each generated object class contains an internal array of triples belonging to it. Class instances are identified by object ID. A KP developer cannot directly create an instance of a class; it accesses the static getInstance() method.

Local storage for all class instances is in the SubscribeQuery class. When you create an instance of a class or receive triples in the notification, the local storage is updated.

To track the need for a class instance in the KP, we use built-in Java SoftReference objects. If there are no references to an instance of the Java class, the Garbage Collector automatically deletes data from the memory after a set time and the SoftReference returns to null. Thus, a reasonable use of resources in the KP is achieved.

Below is the sample code for obtaining an instance of a class "$CLASS_NAME$":

```
public static $CLASS_NAME$ getInstance(String objectId, String accessPointName) {

    $CLASS_NAME$ ret = ($CLASS_NAME$) SubscribeQuery.getInstance().getKnownObject(objectId);
```

```
        if (ret == null) {
            ret = new $CLASS_NAME$(objectId, accessPointName);
            SubscribeQuery.getInstance().registerObject(ret);
        }
        return ret;
    }
```

As a result, we get a two-step model of KP development: 1) code generation based on ontology; and 2) KP implementation based on the generated code. If there are no manual changes in the generated code, the ontology can be extended without affecting the control. The resulting code contains the required features for working with the SIB: obtaining objects; tracking modifications; and ensuring data consistency.

Conclusion

This chapter discusses the problem of using an object model to implement the interaction of software agents based on the Smart-M3 platform. Using a single implementation template allows objects based on a set of triples to be built and using smart notifications to receive object changes. The approach to using the centralized object storage model allows maintenance of data consistency and minimizes the load on the SIB.

The approach to code generation based on the ontology model has its advantages and disadvantages. The advantages include that it is open source and contains all the features of working with SIB. As a disadvantage, a reduction in the flexibility of the development of software agents can be noted.

Acknowledgment

This research was financially supported by the Russian Fund for Basic Research through research project # 19-07-01027. The results were implemented by the Government Program of Flagship University Development for Petrozavodsk State University in 2017–2021.

References

Batanov, D. N. and W. Vongdoiwang. 2007. "Using Ontologies to Create Object Model for Object-Oriented Software Engineering". *Springer US*: 461–487. https://doi.org/10.1007/978-0-387-37022-4_16

Brickley, D. 2004. "Basic Geo (WGS84 lat/long) Vocabulary". *W3C Semantic Web Interest Group*. https://www.w3.org/2003/01/geo/

Catania, V. and D. Ventura. 2014. "An approach for monitoring and smart planning of urban solid waste management using smart-m3 platform". *Proceedings of 15th Conference of Open Innovations Association FRUCT*: 24–31.

Davies, J., R. Studer, and P. Warren. 2006. *Semantic Web technologies: trends and research in ontology-based systems*. John Wiley & Sons.

Evermann, J. and Y. Wand. 2005. "Ontology based object-oriented domain modelling: fundamental concepts" *Requirements engineering* 10: 146–160.

Galov, I. and D. Korzun. 2014. "A notification model for Smart-M3 applications". *Proc. 14th Int'l Conf. Next Generation Wired/Wireless Networking and 7th Conf. on Internet of Things and Smart Spaces (NEW2AN/ruSMART 2014)*: 121–132.

Hannemann, J. and G. Kiczales. 2002. Design pattern implementation in Java and AspectJ. *ACM Sigplan Notices* 37(11): 161-173.

Honkola, J., H. Laine, R. Brown, and O. Tyrkkö. 2010. "Smart-M3 information sharing platform". *Proc. IEEE Symp. Computers and Communications (ISCC'10)*: 1041–1046.

Knublauch, H. 2004. "Ontology-driven software development in the context of the semantic web: An example scenario with Protege/OWL". *1st International workshop on the model-driven semantic web (MDSW2004)*: 381–401.

Korzun, D., A. Lomov, P. Vanag, J. Honkola, and S. Balandin. 2010. "Generating modest high-level ontology libraries for Smart-M3" *Proc. 4th Int'l Conf. Mobile Ubiquitous Computing, Systems, Services and Technologies (UBICOMM 2010)*: 103–109.

Korzun, D. G., S. I. Balandin, V. Luukkala, P. Liuha, and A. V. Gurtov. 2011. "Overview of Smart-M3 principles for application development". *Proc. Congress on Information Systems and Technologies (IS&IT'11), Conf. Artificial Intelligence and Systems (AIS'11)* 4: 64–71.

Kulakov, K. and O. Petrina. 2015. "Ontological model of multi-source smart space content for use in cultural heritage trip planning". *Proc. 17th Conf. Open Innovations Framework Program FRUCT*: 96–103.

Kulakov, K., S. Marchenkov, and S. Tishkov. 2019. "An approach to generating ontology-based object model for Smart-M3". *Proceedings of the 24th Conference of Open Innovations Association FRUCT*: 670–676.

Lomov, A. A. 2013. "Ontology-based KP development for Smart-M3 applications". *Proc. 13th Conf. of Open Innovations Association FRUCT and 2nd Seminar on e-Tourism for Karelia and Oulu Region*: 94–100.

Vdovenko, A. and D. Korzun. 2014. "Planning social activity in SmartRoom: Ontology-based service design". *Proc. 15th Conf. Open Innovations Framework Program FRUCT*: 221–222.

Vdovenko, A., S. Marchenkov, and D. Korzun. 2015. "Enhancing the smartroom system with e-tourism services". *Proc. 17th Conf. Open Innovations Framework Program FRUCT*: 237–246.

Vdovenko, A. S., D. G. Korzun, and I. V. Galov. 2017. "Simulation performance evaluation of Smart-M3 applications for Internet of Things environments". *Intelligent Data Acquisition and Advanced Computing Systems: Technology and Applications (IDAACS), 9th IEEE International Conference*: 994–999.

Volkel, M. and Y. Sure. 2005. "Rdfreactor – from ontologies to programmatic data access" *Poster Proceedings of the Fourth International Semantic Web Conference.*

CHAPTER 2.5

AN ONTOLOGY-BASED MULTILEVEL INTELLIGENT ASSISTANT FOR TIME SERIES CLASSIFICATION

MAN TIANXING[1,*], NATALY ZHUKOVA[2,3], NIKOLAY G. MUSTAFIN[3]
[1] ITMO University, St. Petersburg, Russia
[2] St. Petersburg Institute for Informatics and Automation of the Russian Academy of Sciences (SPIIRAS), St. Petersburg, Russia
[3] St. Petersburg State Electrotechnical University "LETI", St. Petersburg, Russia
*mantx626@gmail.com, nazhukova@mail.ru, ngmustafin@etu.ru

Abstract. Knowledge-based systems can be used to solve data processing problems. This chapter selects the time series classification problem, which is of concern to data researchers in the field of data processing, as an example of the application of knowledge-based systems. Time series classification (TSC) is a typical supervised learning problem, which has been applied widely in sensor data analysis. We built a multilevel intelligent assistant to help users solve the TSC problem. The assistant uses knowledge of the domain of data processing presented in the form of an ontology. When the user inputs the characteristics of the data and task requirements, they can generate reasonable suggestions and a description of the workflow of the algorithm. The data characteristics are separated into multiple levels based on the state during processing: raw data; integrated data; indicator data; tidy data; and output model. Algorithms at different levels process the corresponding data forms until the appropriate output model is generated. Experiments show that knowledge-based intelligent systems can guide the construction of a process for data processing by describing domain knowledge, while the multilevel structure reduces the complexity of data processing.

Introduction

A knowledge-based system for data processing is a type of recommendation system. It recommends items meeting a user's requirements in an interactive manner. At the request of the user, the corresponding solution is generated (Korzun & Meigal, 2019). As an example of the application of a knowledge-based recommendation system, this chapter presents an ontology-based intelligent assistant for solving the time series classification problem. Its core is made up of descriptions of the details and features of time series processing-related knowledge. The core is built using ontological technology.

Time series (TS) data analysis is an important topic in the field of machine learning and data mining. The time series classification (TSC) problem is a classic problem of time series analysis (Xing et al. 2011). TSC is currently being applied in many different fields (Rakthanmanon et al. 2012).

Many researchers are working on reviews of the advantages and disadvantages of different TSC algorithms. We have previously presented a knowledge-based recommendation system for time series classification with the purpose of making such overviews more user-friendly. Knowledge about algorithms, data features, output requirements, etc., which are related to the process of TSC, are described in the ontology. Object properties and data properties are used to describe the logic relations between algorithms and task requirements (Tianxing et al. 2019). Such representations of knowledge can help users understand relevant TSC knowledge. With the help of this system, users can discover suitable solutions for distinct data processing tasks according to the characteristics of input data and task requirements.

The processing of time series data is a complex process. The final output model is generated after multiple data conversions. The complexity of the problem of sorting out the entire process has not yet been resolved.

Based on the ontology used in the proposed ontology-based recommendation system, we define a new ontology that structurally optimizes the description of the knowledge. The ontology has a multilevel structure. Levels contain the characteristics of the processed data corresponding to the different states of the data. Such a structure greatly reduces the complexity of the construction of the data processing processes. This is different to the traditional way of describing data on a single level. The multilevel structure makes users consider only the conditions and solutions corresponding to the relevant level.

The main advantages of the multilevel intelligent assistant as a knowledge-based system are:

- It contains most of the TSC algorithms and data preprocessing methods and provides measurement and representation selection for some of these algorithms.
- It is based on ontological technology. An important feature of ontology is that it can dynamically import and extend any data model by directly integrating various ontological models. The system is easily extended when a new TSC algorithm is proposed or a new data feature considered.
- It is different from other existing recommendation systems that only consider classification or clustering. The ontology also describes time series preprocessing and representation methods.
- It creates abstract classes to describe the details of each method based on certain defined properties so that users can better understand the process of data processing.

The Time Series Representation Problem

In order to effectively speed up the processing of time series data, different representations need to be built. How to best represent a time series is the basic problem of time series mining (Xing et al. 2010). An effective time series representation method does not just allow comparison of the similarity between sequences, but can also be applied to different data mining tasks (Abanda et al. 2018).

Taking into account different conversion methods, Ratanamahatana et al. (2005) classified time series representations into three types: non-data adaptive, data adaptive, and model based.

Non-data Adaptive

In the non-data adaptive representation method, the conversion parameters of each time series are consistent. Spectrum analysis is a relatively common non-data adaptive representation. Agrawal, Rakesh et al. (1993, 69) used the discrete Fourier transform (DFT) to map time series to the frequency domain. This method effectively solves the two problems of "completeness of feature extraction" and "dimension disaster" in time series mining. Chan proposed the use of a discrete wavelet transform (DWT) to process time series data (Chan & Fu, 1999).

In addition to spectrum analysis, other researchers have proposed other methods specifically for time series representation. For example, Keogh and Eamonn et al. (1998, 263) proposed a method based on piecewise linear segments to represent the shape of a time series. This method is suitable for rapid time series classification and clustering. Later, Keogh et al. (2001) proposed the new dimension reduction technique of piecewise aggregate approximation (PAA). They also described the advantages that PAA provides in time series similarity measure and index.

Data Adaptive

A data adaptive representation assumes that at the time of data conversion, the conversion parameters change as the time series data changes. Non-data adaptive representation methods can be transformed into data adaptive representation methods. For example, Keogh et al. (2001, 236) proposed an adaptive piecewise constant approximation (APCA) for time series based on PAA.

Lin et al. (2007) proposed a symbolic aggregate approximation (SAX) representation method that converts initial real-valued high-dimensional data into discrete low-dimensional data.

Ye and Keogh (2009) proposed the concept of time series shapelets as a sub-sequence that best represents a category.

Model-based

A model-based representation assumes that a time series is an observation of a potential model. Azzouzi, Mehdi, and Nabney (1998) first proposed the use of a hidden Markov model (HMM) to define the relationship between time series variables. Kalpakis et al. (2001) used the auto-regression integrated moving average (ARIMA) to succinctly represent time series data; they defined a highly efficient similarity measure for this method. Nanopoulos et al. (2001) proposed a feature extraction method based on statistical models (such as mean, variance etc.) to represent the entire time series.

Generally, a model-based representation has strong interpretability. Two time series may be represented by the same potential model of the same parameter set and be considered similar (Susto et al. 2018).

The Time Series Classification Problem

The TSC problem is defined as follows. Given a data sample set D = $\{(X_1, c), (X_2, c) \dots (X_n, c)\}$, and its discrete labels C = $\{c_1, c_2 \dots c_k\}$, the timestamps are Y = $\{y_1, y_2 \dots y_m\}$. Each data sample X_i includes: an n-dimensional time vector and an associated category c : $X_i = \{X_i^1, X_i^2 \dots . X_i^n, c\}$. TSC trains a classifier on a dataset D in order to map from the space of possible inputs to a probability distribution over the class variable values.

Due to the particularity of time series data, the problem of time series classification faces three major challenges. First, for most classifiers such as decision trees or neural networks, the input data is a feature vector; however, time series data has no clear features. Secondly, although the feature selection method can be used on a time series, due to the dimension of the feature space being very large, the process of feature selection is very cumbersome, and the quantity of calculation required is very large; finally, in some applications, in addition to accurate classification results, the user also wishes to obtain an interpretable classifier. However, since time series data has no clear characteristics, it is very difficult to establish an interpretable classifier.

All classification problems depend on similarity between data, and the time series classification problem is no exception. The similarity between time series has the following three categories (Bagnall et al. 2012).

Similarity in Time

Time series of the same category present the results of observing a potentially identical curve in the time dimension, and the difference between them may be caused by noise and phase drift. The 1-NN classifier is best suited to handle such problems, while the DTW metric mitigates the effects of noise.

Similarity in Shape

Time series of the same category are distinguished by some identical sub-sequences or shapes, which may appear anywhere in the time series. This is the main difference to the similarity in the time domain. The smaller the correlation of sub-sequences with time, the more difficult it is for time domain-based 1-NN classifiers to deal with such problems. In this case, different categories can be distinguished by using time series feature-based methods.

Similarity in Change

Such similarity is difficult to observe appears in a highly auto-correlated sequence. This problem can be handled in a production mode, such as a hidden Markov model (HMM), an auto-regressive moving average (ARMA), and so on (Tuncel & Baydogan 2018).

Existing Overviews of TSC

It is difficult for a non-computer professional to choose a method that meets their needs. Many scholars are committed to presenting a comprehensive and understandable TSC taxonomy. Bagnall et al. (2017) implemented common TSC methods and compared their performance in a large number of experiments. However, they do not deal with model-based algorithms. Fortunately, Fawaz et al. (2019) present a review of deep learning for TSC and a detailed description and comparison of the model-based TSC algorithm. The experiments of these two reviews are based on a univariate TSC benchmark (the UCR/UEA archive; Dau et al. 2018; Bagnall et al. 2018) and so this article can summarize the performance of most TSC algorithms according to the results they produce (Sirisambhand & Ratanamahatana 2019).

Concurrently, Bagnall et al. (2017) reviewed the common representations of time series to provide a basis for research into TSC. Lhermitte et al. (2011) present a comparison of similarity measures for major time series.

Ontology Technology

Ontology can be considered a modelling tool to describe domain concepts at semantic and knowledge levels. Ontology aims to capture knowledge in related fields, identify common recognized terms in the domain, describe the semantics of concepts through the relations between them, and provide a common understanding of knowledge in a field. The proposed intelligent assistant is primarily based on ontology technology. Its main feature is to make information on the Web have comprehensible semantics for computers, realize semantic interoperability between information systems with ontological support, and intelligently access and retrieve Web resources. As a result, the assistant displays good scalability (Korzun, Kashevnik, & Balandin 2017). In the future we will be able to connect the current basic assistant to other existing ontologies and so extend it.

Ontological technology is used as a method to implement the assistant because of the advantages it offers.

- In ontology, in addition to the relation "has-a", we can define more relationships to represent the links between algorithms and the characteristics of a dataset. Flexible and clear relationships allow users to find the correct solution more accurately and quickly.
- The implemented ontology presents a more complex mind map to the user and the user does not need to understand any computer language. Each input can get the relevant information according to the relationships in the ontology, making it suitable for non-computer professional researchers.
- To solve data classification problems, an excess of knowledge is involved. The intelligent assistant deals with 45 classification algorithms, with different descriptions of the processes. The performance of the algorithms has been estimated according to a variety of datasets with different attributes. A flowchart cannot describe such a quantity of knowledge and a general taxonomy cannot describe such complex information.

Multilevel Structure for Ontology

Ontology has recently been introduced as a formal structure to describe things, attributes, relations, and general domain theories. It is used in the fields of artificial intelligence, information retrieval, and knowledge engineering.

The support of ontology-based conceptual modelling mainly focused on an analysis of the properties of individual categories. As a consequence, these approaches are unable to describe the evolution of a hierarchy of individual properties and the processing of TSC involves the process of the multi-level evolution of TS data.

The need to support the representation of subject domains, such as TS processing dealing with multiple levels, has given rise to what has been referred to as multilevel modelling.

Pai, Yang, and Chung (2017) present a multi-layer ontology-based information fusion for situation awareness. We use a similar idea to describe knowledge about the TSC problem. Carvalho (2016) presents the main method to build a multilayer ontology conceptual model. The method shows that we can transform a thing from one level to another by meeting special conditions.

Ontology-based Multilevel Intelligent Assistant

We built this ontology-based multilevel intelligent assistant using overviews of existing TSC algorithms. Building an ontology assumes the creation of entities associated with TSC and the definition of new properties to represent the connections between them. Based on the characteristics of the algorithms and a large number of experiments, we can link the characteristics of the dataset to suitable algorithms, measures, and representations.

The overall conceptual model of the ontology is presented in Fig. 1.

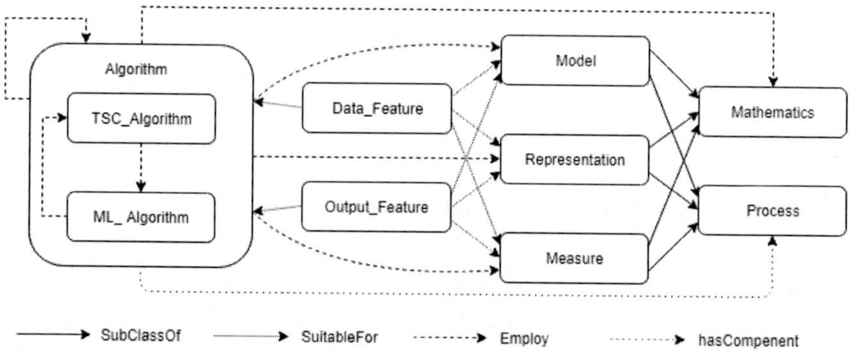

Figure 1. The conceptual model of the ontology

Bagnall et al. (2017) did a lot of experiments comparing the performance of most TSC algorithms. The work of Fawaz et al. (2019) is mainly devoted to the application of deep learning in TSC. Based on these experimental results, we summarized the common characteristics of datasets and user requirements and linked them to algorithms that can be used for data processing. This is the basis for how the assistant provides advice on choosing an algorithm.

We used classes of representation methods, measures, mathematics, and processes. Processes describe the logic of data processing.

In order to make the data processing process clearer, the intelligent assistant has a multilevel structure that reflects the main steps of data processing processes. The structure of the process of data processing is given in Fig. 2.

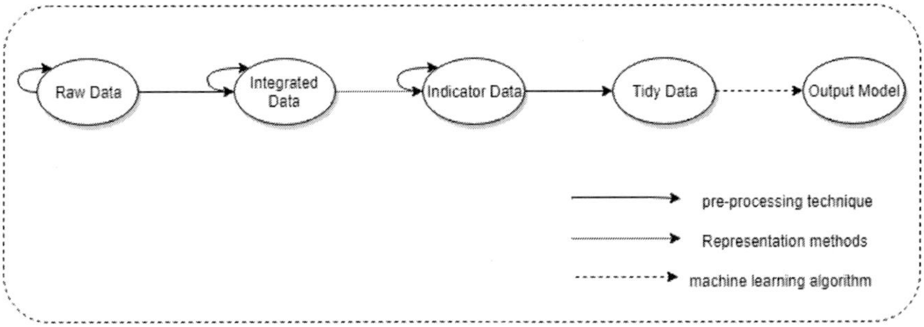

Figure 2. The structure of the data processing process

Raw data comprises the initial datasets collected from the sources. Usually these data are stored in different forms and the sets include missing values, noise and so on. (Korde & Mahender 2012, 85).

Integrated data is generated from raw data through some basic preprocessing operations. But at this level, data is still presented as a set of separate data elements.

Indicator data is the form that can be used to describe information contained in the initial data. For better classification results, some data defects can be fixed to adapt the data to the classification algorithms.

Tidy data is the data form that data processing can deal with. We use some entities to describe the current data characteristics as the basis for algorithm selection. Data is connected to the algorithms suitable for its classification in different situations by the object property.

Output model comprises the results of TSC tasks. In addition to data characteristics, we consider the task requirements to choose the appropriate algorithms to generate the output model. These requirements are described in the "output model" level.

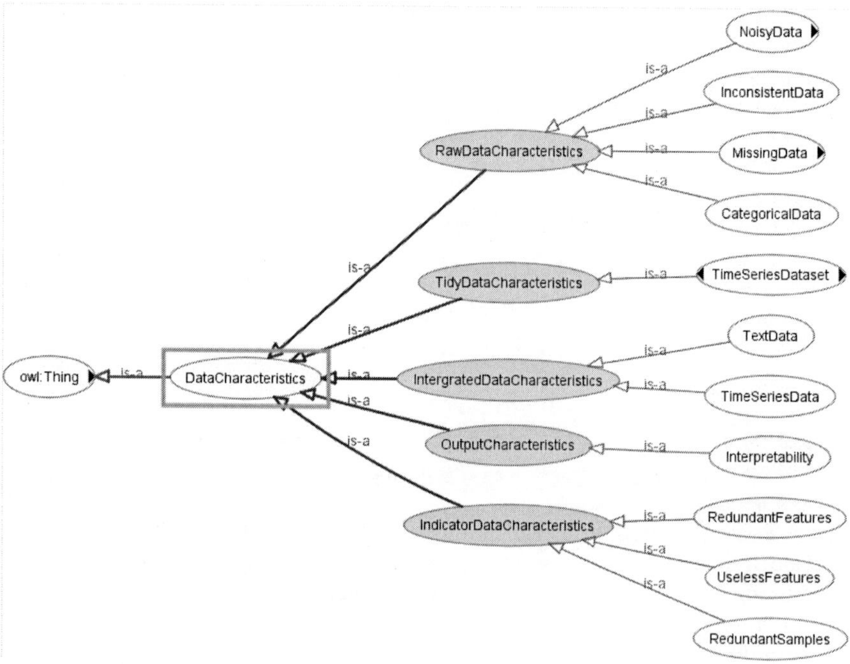

Figure 3. Examples of data characteristics at each level

As Fig. 3 shows, the data at each level has unique characteristics and presents problems to be solved. The levels are defined with the data property "inLevel" as the form "DataCharacteristics and inLevel exactly *n* xsd:int". "DataCharacteristics" is the super class while "inLevel" is the object property pointing to the number of the corresponding level (*n*).

The advantage of this multilevel structure is that at each level only the problems that correspond to the specified level are considered and irrelevant characteristics cannot be seen. With transformation between data levels, a reasonable operation process is generated.

Data processing algorithms should be applied in the correct sequence. Some repetitive operations can be applied without consideration of the state of the data.

After acquiring the dataset to be processed, the characteristics of the data are evaluated at the lowest level. When all processing difficulties caused by the characteristics of the data at the current level are solved, the data is then converted to the next level. In general data processing problems, data conversion can be done only through preprocessing techniques. But the particularity of time-series makes it necessary to use methods for building data representations before classification algorithms can be applied.

As Fig. 2 shows, preprocessing techniques are used for conversion within the levels of raw data, integrated data, and indicator data. Data is also converted from raw data to integrated data and from indicator data to tidy data by preprocessing techniques. The essence of these conversion operations is to eliminate data characteristics that are detrimental to the final classification task.

Representation methods are used to convert the time series data into the data form required by the classification algorithm. In the ontology they are described as operations of converting integrated data to indicator data.

The conversion from tidy data to the output model is implemented by TSC algorithms. The data are processed by the classification algorithm to generate a corresponding output model. The characteristics of the tidy data and the requirements of the output model are the basis for choosing appropriate TSC algorithms.

To ensure the rationality and logic of the ontology, classes and properties make up the core content. The following classes are defined in the ontology.

Algorithm. Algorithms for data processing can be common machine learning algorithms, such as the ML_Algorithm (Tianxing & Zhukova, 2018), and special algorithms oriented to time series classification (as shown in Table 1). The main TSC algorithms are shown in Fig. 4.

Figure 4. TSC algorithms and a parent class "AlgorithmBasedOnShapelets" in the ontology

Table 1. List of TSC Algorithms

No.	Name in ontology	Full Name
1	ACF	Autocorrelation function
2	BN	bayesian network
3	BoP	Bag of Patterns
4	BOSS	Bag of SFA Symbols
5	C45	C4.5
6	CID_DTW	Complexity-Invariant Distance
7	COTE	Collection of Transformation E
8	DD_DTW	Derivative Dynamic Time Warping
9	DDTW_R1_1NN	Derivative Dynamic Time Warping with full warping window

10	DDTW_Rn_1NN	Derivative Dynamic Time Warping with warping window set through cross validation
11	DTD_C	Derivative Transform Distance
12	DTW_F	Dynamic Time Warping Features
13	DTW_R1_1NN	Dynamic Time Warping with full warping window
14	DTW_Rn_1NN	Dynamic Time Warpingwith warping window set through cross validation
15	EE	Elastic Ensemble
16	Encoder	Encoder
17	ERP_1NN	Edit Distance for Real Sequences_1-nearest neighbor algorithm
18	Euclidean_1NN	Euclidean_1-nearest neighbor algorithm
19	FCN	Fully Convolutional Neural Network
20	FS	Fast Shapelet Tree
21	LCSS_1NN	Longest Common Subsequence_1-nearest neighbor algorithm
22	Logistic	Logistic regression
23	LPS	Learned Pattern Similarity
24	LS	Learned Shapelets
25	MCDCNN	Multi-Channel Deep Convolutional Neural Network
26	MCNN	Multi-scale Convolutional Neural Network
27	MLP	Multi-layer Perceptron
28	MSM_1NN	Move-Split-Merge
29	NB	Naïve Bayes
30	PS	power spectrum transform
31	RandF	Random_Forest
32	ResNet	Residual Network
33	RotF	Rotation_Forest
34	SAXVSM	Symbolic Aggregate approximation and Vector Space Model
35	ST	Shapelet Transform
36	SVML	Support Vector Machine with linear kernel
37	SVMQ	Support Vector Machine with quadratic kernel
38	Time-CNN	Time Convolutional Neural Network
39	t-LeNet	Time Le-Net
40	TSBF	Time Series Bag of Features
41	TSF	Time Series Forest
42	TWE_1NN	Time Warp Edit Distance
43	TWIESN	Time Warping Invariant Echo State Network
44	WDDTW_1NN	Weighted Derivative Dynamic Time Warping
45	WDTW_1NN	Weighted Dynamic Time Warping

Data_Feature. Characteristics of input datasets, including size, length etc. We define the description with a data property "inLevel" to present the multilevel process of data transformation. For example, the characteristics of the raw data are described as "DataCharacteristics and inLevel exactly *0* xsd:int".

Output_Feature. Features of the output data and user requirements. This is the 4th level of data representation, described as "DataCharacteristics and inLevel exactly *4* xsd:int".

Mathematics. The mathematical base of the algorithms.

Measure. Measures for estimating expected and actual results of data processing and corresponding similarity functions.

Model. Basic algorithm models.

Representation. Common time series representation methods. We summarize the comparison of these methods in Table 2 and describe them in the ontology. The differences of these methods are also described in the ontology. Sometimes users can consider these comparisons as conditions for choosing suitable algorithms.

Process. An abstract concept to make the description of the algorithm clearer, containing a description of the mathematics, algorithm, and measure. They link the algorithms that contain them with the relationship "hasComponent".

Table 2. Comparison of Time Series Representation

(*Columns in table: 1-Time domain frequency domain transform; 2-Dimensionality reduction; 3-Linear computational complexity; 4-Symbolization; 5-Processing variable length sequences; 6-Dynamic insertion/deletion; 7-Understandable; 8-Maintaining local feature*)

Representation Method	**1**	**2**	**3**	**4**	**5**	**6**	**7**	**8**
DFT	√	√	×	×	√	×	×	×
DWT	√	√	×	×	×	×	×	×
SVD	√	√	×	×	√	×	×	×
PAA	×	√	√	×	√	√	√	×
APCA	×	√	×	×	√	√	√	×
PLA	×	√	×	×	√	√	√	×

PRA	×	√	√	×	√	√	×	×
Polynomial fitting	×	√	×	×	√	×	×	×
Clipper Data	×	×	√	√	√	√	√	×
SAX	×	√	√	√	√	√	√	×
Landmarks	×	√	√	×	√	√	√	√
Important point	×	√	√	×	√	√	√	√

Object properties make ontology more flexible than taxonomy. We define some necessary properties to describe the relationship between classes.

employ. A property that links the class “algorithm” with “measure”, “representation” and “mathematics”. If an algorithm consists of a number of algorithms, then the property links these algorithms.

hasComponent. Links the algorithms considering the data processing steps.

suitableFor. Links the input information and the algorithms, defining suitable algorithms for processing data with known characteristics.

hasSize. A data property in ontology is used to describe the value and range of the datasets. Part of the description with data property “hasSize” in the ontology are shown in Table 3.

inLevel. Another data property in the ontology that is used to describe different characteristics of the data at different levels.

Table 3. Part of the Descriptions with Data Property “hasSize” in the Ontology

Category	Range of Class Value	Entity in Ontology
Sample size	hasSize some xsd:integer[< 100]	SmallTrainTSDataset
	hasSize some xsd:integer[>= 100, <=500]	SmallTrainTSDataset
	hasSize some xsd:integer[> 500]	LargeTrainTSDataset
Attribute size	hasSize some xsd:integer[< 300]	ShortTSDataset
	hasSize some xsd:integer[>= 300, <=700]	MediumTSDataset
	hasSize some xsd:integer[> 700]	LongTSDataset
No. of classes	hasSize some xsd:integer[< 10]	FewClassTSDataset
	hasSize some xsd:integer[>= 10, <=30]	MediumClassTSDataset
	hasSize some xsd:integer[> 30]	ManyClassTSDataset

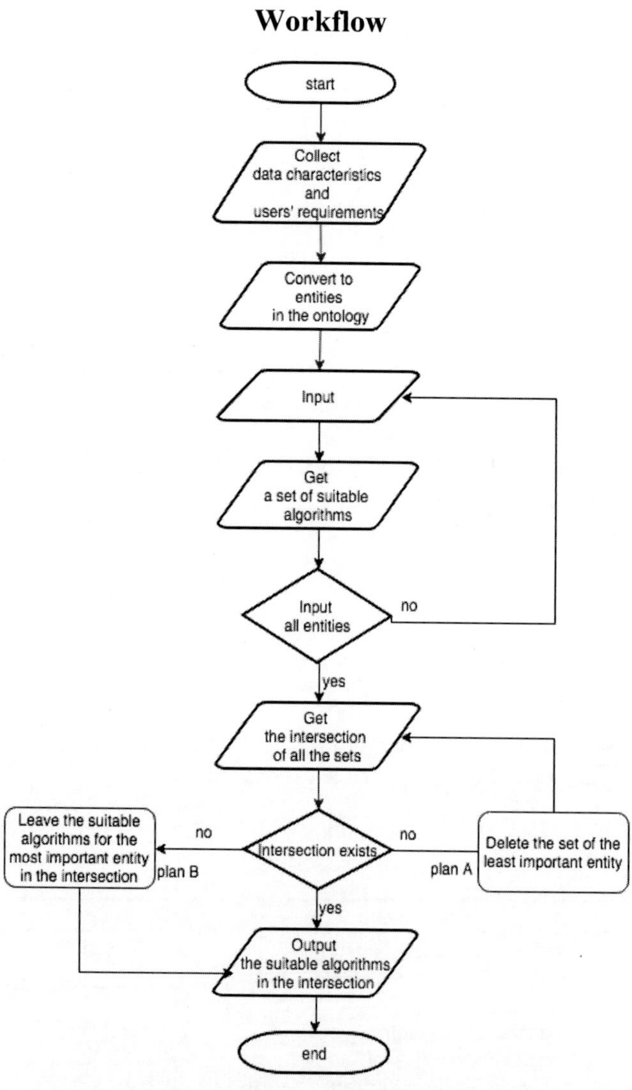

Figure 5. The workflow of the intelligent assistant

The main goal of the intelligent assistant is to choose suitable solutions for the TSC task. The whole process of solution selection consists of four steps. The flowchart is shown in Fig. 5:

1) In the current level, collect the characteristics of datasets and the requirements of the task and describe them with the ontological entities in the corresponding level.
2) Input each of the characteristics and requirements and get the sets of suitable algorithms.
3) Find the best choice among the suitable algorithms based on the user conditions (generally, we extract the intersection of these sets).
4) (Optional) Describe the selected algorithm based on the outward links.

In this way the final result is a reasonable data analysis process. The original dataset goes through a continuous conversion process: starting out disordered and gradually becoming regular. Finally, a TSC algorithm is applied to process the dataset, generating an efficient classifier. his can provide users with relevant conclusions of data analysis.

It is worth noting that sometimes the user's needs are so difficult to meet that no algorithm can satisfy all the required conditions. There are two solutions depending on the needs of the user:

1) Delete the conditions that the user thinks are least important, and then find the intersection.

2) Select the algorithms that are appropriate for the conditions that the user considers to be the most important to match the candidate algorithms.

Experiment

The suggested intelligent assistant was used to process a classical TS dataset "CinCECGtorso". This dataset is derived from a Computers in Cardiology challenge, an annual competition that runs with a conference series of the same name and is hosted on PhysioNet. It uses ECG data from multiple torso-surface sites. There are four classes (four different people).

First, the characteristics of this dataset were analyzed and described with the entities in the ontology. In this case, the data set has been preprocessed, so the state of the current data is in the level "TidyCharacteristics" and we should consider the corresponding entities, as shown in Fig. 6.

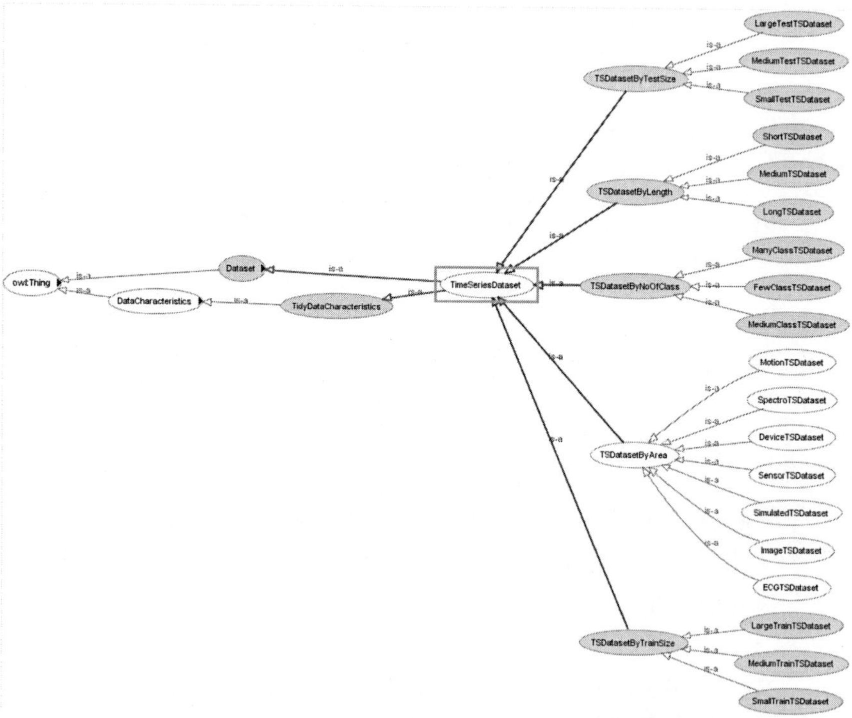

Figure 6. TS characteristics in the level "TidyCharacteristics"

The corresponding entities of the dataset characteristics are shown in Table 4.

Table 4. The Corresponding Entities of the Dataset Characteristics

Category	Value of dataset	Range of Class Value	Entities
Train size	40	hasSize some xsd:integer[< 100]	SmallTrainTSDataset
Test size	1380	hasSize some xsd:integer[> 1000]	LargeTestTSDataset
Length	1639	hasSize some xsd:integer[> 700]	LongTSDataset
No. of classes	4	hasSize some xsd:integer[< 10]	FewClassTSDataset
Data area	ECG	ECG	ECGTSDataset

The results of the selection of all the suitable algorithms for CinCECGtorso are shown in Table 5.

Table 5. All Suitable Algorithms for CinCECGtorso

(*The symbol √ means this algorithm is suitable for the dataset in this category. The highlighted algorithms are those that are suitable for all the requirements*)

Algorithm	Train size	Test size	Length	No. of classes	Data area
BOSS	√	√	√	√	√
CID_DTW	√	√	√		
COTE	√	√	√	√	√
DD_DTW			√		
DTD_C			√		
DTW_F	√	√	√	√	
EE	√	√	√	√	√
ERP_1NN					√
LCSS_1NN					√
LPS	√	√		√	
LS	√	√		√	
MSM_1NN	√	√	√	√	√
PS					√
ST	√	√	√	√	√
SVMQ					√
TSBF	√	√		√	
TSF			√	√	√

In this experiment, BOSS, COTE, EE, MSM_1NN and ST were selected for the dataset CinCECGtorso, since these algorithms are suitable for all relevant conditions, as shown in Fig. 7.

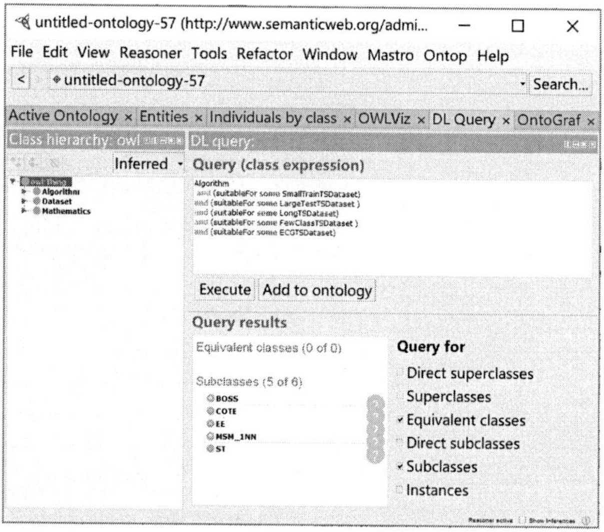

Figure 7. Algorithms suitable for the dataset CinCECGtorso

We applied all the TSC algorithms on the dataset. The accuracy of all the TSC algorithms is ranked in Table 6.

Table 6. The Accuracy Rank of All TSC Algorithms for CinCECGtorso

Rank	Algorithm	Accuracy	Rank	Algorithm	Accuracy
1	COTE	0.983	24	LPS	0.743
2	TSF	0.974	25	FS	0.741
3	CID_DTW	0.954	26	DD_DTW	0.731
4	EE	0.946	27	RandF	0.731
5	DDTW_Rn_1NN	0.944	28	SAXVSM	0.730
6	WDDTW_1NN	0.938	29	DDTW_R1_1NN	0.717
7	MSM_1NN	0.935	30	TSBF	0.716
8	LCSS_1NN	0.928	31	BoP	0.716
9	DTW_Rn_1NN	0.928	32	DTW_F	0.714
10	ST	0.918	33	RotF	0.712
11	WDTW_1NN	0.908	34	DTW_R1_1NN	0.674
12	BOSS	0.900	35	SVMQ	0.657
13	ERP_1NN	0.899	36	MCDCNN	0.643
14	Euclidean_1NN	0.891	37	C45	0.604
15	PS	0.888	38	Time-CNN	0.600
16	LS	0.855	39	Encoder	0.573

17	NB	0.847	40	TWIESN	0.553
18	TWE_1NN	0.846	41	MCNN	0.533
19	ResNet	0.844	42	t-LeNet	0.533
20	DTD_C	0.820	43	SVML	0.462
21	FCN	0.814	44	MLP	0.462
22	BN	0.803	45	Logistic	0.379
23	ACF	0.786			

As Table 6 shows, the chosen algorithms clearly lie in the upper half and show good performance. At least we can say that with the help of the intelligent assistant we have not made bad choices.

Conclusion

This chapter presents an ontology-based multilevel intelligent assistant for solving the time series classification problem. Such TSC problems are described in the form of an ontology with corresponding classes and properties. Description of the structure of the data characteristics at multiple levels improves the efficiency of the proposed recommendation system.

As a typical knowledge-based system for TSC, the ontology-based multilevel intelligent assistant has greater flexibility than other existing taxonomies. Such an intelligent assistant can effectively help non-experts to build and understand an appropriate data processing processes oriented towards TSC.

In general, the efficiency of the assistant also proves the good application prospects of knowledge-based systems in the field of data processing.

Bibliography

Xing, Z., Pei, J., & Keogh, E. (2010). A brief survey on sequence classification. *ACM Sigkdd Explorations Newsletter, 12(*1), 40-48.

Rakthanmanon, T., Campana, B., Mueen, A., Batista, G., Westover, B., Zhu, Q., & Keogh, E. (2012, August). Searching and mining trillions of time series subsequences under dynamic time warping. In *Proceedings of the 18th ACM SIGKDD international conference on Knowledge discovery and data mining* (pp. 262-270). ACM.

Ratanamahatana, C., Keogh, E., Bagnall, A. J., & Lonardi, S. (2005, May). A novel bit level time series representation with implication of similarity search and clustering. In *Pacific-Asia Conference on Knowledge Discovery and Data Mining* (pp. 771-777). Springer, Berlin, Heidelberg.

Agrawal, R., Faloutsos, C., & Swami, A. (1993, October). Efficient similarity search in sequence databases. In *International conference on foundations of data organization and algorithms* (pp. 69-84). Springer, Berlin, Heidelberg.

Chan, K. P., & Fu, A. W. C. (1999, March). Efficient time series matching by wavelets. In *Proceedings 15th International Conference on Data Engineering (Cat. No. 99CB36337)* (pp. 126-133). IEEE.

Keogh, E. J., & Pazzani, M. J. (1998, August). An Enhanced Representation of Time Series Which Allows Fast and Accurate Classification, Clustering and Relevance Feedback. In *Kdd* (Vol. 98, pp. 239-243).

Keogh, E., Chakrabarti, K., Pazzani, M., & Mehrotra, S. (2001). Dimensionality reduction for fast similarity search in large time series databases. *Knowledge and information Systems*, *3*(3), 263-286.

Keogh, E., Chu, S., Hart, D., & Pazzani, M. (2001, November). An online algorithm for segmenting time series. In *Proceedings 2001 IEEE International Conference on Data Mining* (pp. 289-296). IEEE.

Lin, J., Keogh, E., Wei, L., & Lonardi, S. (2007). Experiencing SAX: a novel symbolic representation of time series. *Data Mining and knowledge discovery*, *15*(2), 107-144.

Ye, L., & Keogh, E. (2009, June). Time series shapelets: a new primitive for data mining. In *Proceedings of the 15th ACM SIGKDD international conference on Knowledge discovery and data mining* (pp. 947-956). ACM.

Ye, L., & Keogh, E. (2011). Time series shapelets: a novel technique that allows accurate, interpretable and fast classification. *Data mining and knowledge discovery*, *22*(1-2), 149-182.

Azzouzi, M., & Nabney, I. T. (1998, September). Analysing time series structure with Hidden Markov Models. In *Neural Networks for Signal Processing VIII. Proceedings of the 1998 IEEE Signal Processing Society Workshop (Cat. No. 98TH8378)* (pp. 402-408). IEEE.

Kalpakis, K., Gada, D., & Puttagunta, V. (2001, November). Distance measures for effective clustering of ARIMA time-series. In *Proceedings 2001 IEEE international conference on data mining* (pp. 273-280). IEEE.

Nanopoulos, A., Alcock, R., & Manolopoulos, Y. (2001). Feature-based classification of time-series data. *International Journal of Computer Research*, *10*(3), 49-61.

Xing, Z., Pei, J., Yu, P. S., & Wang, K. (2011, April). Extracting interpretable features for early classification on time series. In *Proceedings of the 2011 SIAM International Conference on Data Mining* (pp. 247-258). Society for Industrial and Applied Mathematics.

Bagnall, A., Davis, L., Hills, J., & Lines, J. (2012, April). Transformation based ensembles for time series classification. In *Proceedings of the 2012 SIAM international conference on data mining* (pp. 307-318). Society for Industrial and Applied Mathematics.

Bagnall, A., Lines, J., Bostrom, A., Large, J., & Keogh, E. (2017). The great time series classification bake off: a review and experimental evaluation of recent algorithmic advances. *Data Mining and Knowledge Discovery*, *31*(3), 606-660.

Fawaz, H. I., Forestier, G., Weber, J., Idoumghar, L., & Muller, P. A. (2019). Deep learning for time series classification: a review. *Data Mining and Knowledge Discovery*, *33*(4), 917-963.

Lhermitte, S., Verbesselt, J., Verstraeten, W. W., & Coppin, P. (2011). A comparison of time series similarity measures for classification and change detection of ecosystem dynamics. *Remote sensing of environment*, *115*(12), 3129-3152.

Tianxing, M., & Zhukova, N. (2018). An Ontology of Machine Learning Algorithms for Human Activity Data Processing. *Learning*, *10*, 12.

Tuncel, K. S., & Baydogan, M. G. (2018). Autoregressive forests for multivariate time series modelling. *Pattern Recognition*, *73*, 202-215.

Abanda, A., Mori, U., & Lozano, J. A. (2019). A review on distance based time series classification. *Data Mining and Knowledge Discovery*, *33*(2), 378-412.

Susto, G. A., Cenedese, A., & Terzi, M. (2018). Time-series classification methods: Review and applications to power systems data. In *Big Data Application in Power Systems* (pp. 179-220). Elsevier.

Dau, H. A., Bagnall, A., Kamgar, K., Yeh, C. C. M., Zhu, Y., Gharghabi, S., & Keogh, E. (2018). The ucr time series archive. *arXiv preprint arXiv:1810.07758*.

Bagnall, A., Dau, H. A., Lines, J., Flynn, M., Large, J., Bostrom, A., & Keogh, E. (2018). The UEA multivariate time series classification archive, 2018. *arXiv preprint arXiv:1811.00075*.

Sirisambhand, K., & Ratanamahatana, C. A. (2019). A Dimensionality Reduction Technique for Time Series Classification Using Additive Representation. In *Third International Congress on Information and Communication Technology* (pp. 717-724). Springer, Singapore.

Pai, F. P., Yang, L. J., & Chung, Y. C. (2017). Multi-layer ontology-based information fusion for situation awareness. *Applied Intelligence*, *46*(2), 285-307.

CARVALHO, V. (2016). Foundations for Multi-level Ontology-based Conceptual Modelling.

Korde, V., & Mahender, C. N. (2012). Text classification and classifiers: A survey. International Journal of Artificial Intelligence & Applications, 3(2), 85.

Tianxing, M., Zhukova, N., & Mustafin, N. (2019, April). A Knowledge-based Recommendation System for Time Series Classification. In *Proceedings of the 24th Conference of Open Innovations Association FRUCT* (p. 108). FRUCT Oy.

Korzun, D., Kashevnik, A., & Balandin, S. (Eds.). (2017). *Novel Design and the Applications of Smart-M3 Platform in the Internet of Things: Emerging Research and Opportunities: Emerging Research and Opportunities*. IGI Global.

Korzun, D., & Meigal, A. (2019, April). Multi-Source Data Sensing in Mobile Personalized Healthcare Systems: Semantic Linking and Data Mining. In *2019 24th Conference of Open Innovations Association (FRUCT)* (pp. 187-192). IEEE.

CHAPTER 2.6

INTELLECTUAL MODELING OF THE INDIVIDUAL EDUCATIONAL TRAJECTORIES OF COMPUTER SCIENCE AND ENGINEERING STUDENTS

EVGENIY V. TITOV[1], ALEXEY V. ERPELEV[2], IGOR V. KRASILNIKOV[3], PAVEL N. MIRONOV[4], ALISA V. PIKOVA[5], VLADIMIR L. SIMONOV[6]

[1] Moscow Technical University of Communications and Informatics. Moscow, Russia. titov@mtuci.ru

[2] Russian State Social University. Moscow, Russia. alexey.erpelev@outlook.com

[3] Financial University under the Government of the Russian Federation. Moscow, Russia. reg009@yandex.ru

[4] State Research Institute of Instrument Engineering Joint-Stock Company. Moscow, Russia. apr@gosniip.ru

[5] Russian State Social University. Moscow, Russia. alisa_vp@mail.ru

[6] Russian State Social University. Moscow, Russia. v.simonov@rambler.ru

Abstract. This article discusses practical aspects of the intellectual modeling of the individual educational trajectories of students in the fields of computer science and computer engineering. Students of programming, electronics, and robotics are considered to be in the most dynamically developing fields in the Russian Federation. Classes in these specialisms aim to develop the basic skills of students in artificial intelligence, robotics, and automation. At the same time, professional skills and relevant professional competencies are being upgraded in accordance with the federal state educational standards for higher education.

Introduction

The formation of individual educational trajectories is determined by many factors, among which we find the following (Titov, Simonov, & Erpelev

2019):

- The organization of the educational curriculum;
- The abilities and inclinations of students;
- The influence of the social environment;
- The availability of additional training opportunities, e.g. through technoparks, clubs, sections, and extra classes.

The above-mentioned examples focus on several goals:

- To prepare someone professionally;
- To uncover a student's creative abilities;
- To maintain student health, especially during the period leading up to adulthood.

Let us consider some of these goals.

Federal state educational standards (FGOS) are relevant to the training of professionals in the Russian Federation; over the past few years a number of these educational standards have changed. Universities are in a hurry to reshape their activities, change curricula, and modify teaching formats. For example, enterprises are helping to organize educational departments. In these departments, classes are organized mainly by specialists, i.e. employees of a particular enterprise.

Uncovering the creative abilities of students takes place through the many institutional structures of the university, including student councils, creative teams, and laboratories.

Regarding the preservation of student health, we should note that if a healthy work and rest regime is not observed, then a student's eyesight may deteriorate, their posture may become poor, and they may develop chronic disease in later years or see long-term damage to the hip area due to excessive uninterrupted sitting at a computer. Therefore, at each educational institution working with young people (and especially at universities), sports grounds, stadiums, swimming pools, and mini-training facilities are built on university campuses. This is extremely important for young people and helps them succeed educationally as well.

The concepts of the "individual educational route" and the "individual educational trajectory" do not have a well-established definition in the theory of pedagogy. As part of this article, the term individual educational route should be understood as the method to achieve individual mastery of curricula. At the same time, such a curriculum is compiled on the basis of the general choices and the corresponding ratio between various educational areas and types of educational work (Schaffhauser 2018; Adaptive learning 2017; Ibatullin & Anisimova 2016; Pozharkova, Noskova, & Troyak 2018).

Theoretical Approach to the Formation of an Individual Educational Trajectory

The standard structure of the basic work program of a discipline or course in a specialty usually has the form shown in Fig. 1. At the beginning, the relevant disciplines D0 are indicated; these are necessary for students to start mastering the course material. The output specifies the competencies K0, obtained by students in the course of studying this material. Unfortunately, this structure is completely ineffective when planning individual educational trajectories.

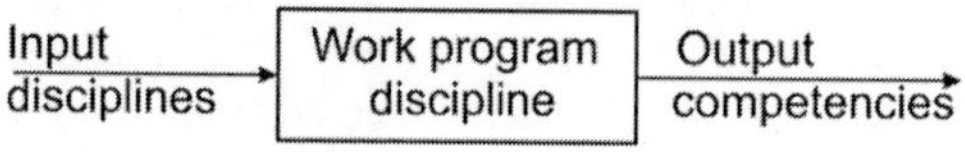

Figure 1. The standard structure of the basic work program of a discipline or course in a specialty

Let us conduct a consistent transformation of the existing structure. In the first step, we replace the input disciplines D0 with the input competencies K1 (Fig. 2), which are outputs for the disciplines D0. At the same time, it is necessary to reduce the list of competencies in order to eliminate duplication. Furthermore, the input and output competencies should not coincide. In addition, the design must acknowledge that not all of output competencies can become input competencies of the studied discipline.

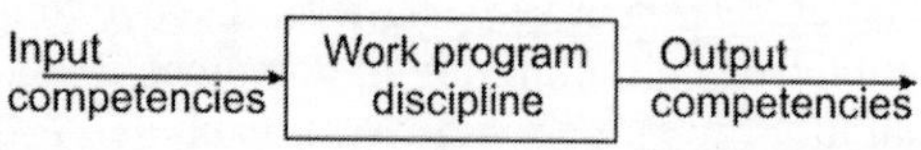

Figure 2. Replacing the input disciplines D0 with the input competencies K1

At the next step, we present a linear structure with an indication of input and output competencies, arranged sequentially within the modules of the discipline under study (Fig. 3). In this scheme, the educational content index (EC) denotes an indivisible module containing educational content. For example, competencies learned during the study of the

Criminal Code must be unambiguously evaluated using the knowledge testing tool T. During a test, each competency is assigned a degree of mastery M. Here, the degree of mastery is estimated on a scale, for example, from 0 to 100 (points or percentage).

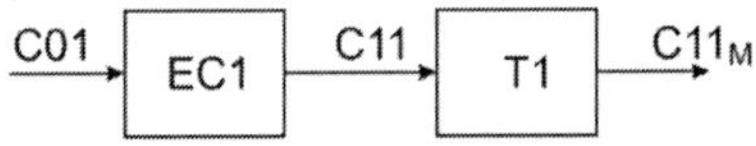

Figure 3. Linear structure with an indication of input and output competencies

However, a number of competencies can be studied within several MCs and evaluated using several Ts. In this case, the degree of mastery for the entire course is assumed to be P = 100 %; for individual modules the maximum degree of mastery is indicated so that these degrees added together give 100 %. For this kind of representation, it is much more efficient to use a representation in the form of a directed graph (Fig. 4).

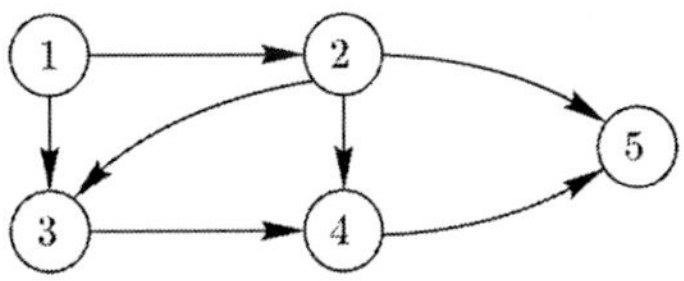

Figure 4. Directed graph

Since the input and output competencies for different training modules may coincide, we can clearly see the options for individual learning trajectories in one course. By changing the amount of material in the training modules, as well as the complexity of the tests, the weights of the development of competencies within a single module can be dynamically changed. As such, the design of educational routes is carried out using tools to find the optimal path on a graph.

A tool such as a graph allows the design of a training course before the start of classes. The use of such a tool is ineffective for dynamically changing the passage of training modules and developing individual educational routes. However, if we replace the static values of the percentage of learning competencies with dynamic weights, then we can convert a static graph into a dynamic neural network.

Since the main task of studying any course is not so much the mastery of specific content, but rather achieving certain competencies, the neural network approach allows the building of an individual educational route, even with the omission of individual training modules optimally adjusted to each student.

Further on in this article specific examples are considered. At the same time we confine ourselves to the training in computer science and computing. In this direction, we consider the fields of programming, electronics, and robotics, being the most dynamically developing areas in the Russian Federation. In connection to the peculiarity of the considered area, the following should be noted.

For the past few decades, the problem of training engineering personnel in the Russian Federation has not been given due attention. This has hindered the development of Russian equipment and technology in comparison to the global level and reduced the prestige of applied engineering, which is unacceptable for the development of the country. In order to handle this situation, several years ago the government of the Russian Federation set out the urgent task of accelerating the development of engineering education and attracting young people to this area. The purpose of these activities is to solve engineering and technological problems and successfully contribute to the development of the country.

In order to intensify the development of engineering education, one of the current modern trends, robotics, was assigned by the Ministry of Education and Science of the Russian Federation to be a priority teaching area in both higher education institutions, schools, and colleges. At the same time, preschool and early school education can also contribute to an increase in the study of the fundamentals of algorithm design, programming, and robotics.

Individual Educational Trajectories of Teaching Programs in the Fields of Artificial Intelligence, Robotics, and Automation

It is important to bear in mind that national education standards, as well as newly integrated professional standards in the field of engineering education, impose strict requirements on university students. These refer to the basics of modern electronics, electrical engineering, practical programming, mechanics, and a number of other related disciplines. Automation system prototypes and robotics are developed on the basis of these disciplines. In accordance with the Federal State Higher Education

Standards, the design of real systems do not focus solely on theoretical knowledge, but also on practical and professional competence.

In particular, it is necessary to have knowledge of the relevant hardware and be able to take part in the creation of software components. Additionally, it is necessary to have sufficient expertise for practical applications in the design of architectural systems; the construction and documentation of software and hardware products; and a willingness to adapt to changing operating conditions (Federal State Educational Standard of Higher Education, bachelor level 2016).

The Russian Ministry of Education has placed a focus on the needs of technological industries in its new education standards. It should be noted that in recent years, professional standards have been quite extensively developed. Interested parties are directly involved in the development of professional standards. As such, various non-profit organizations and companies are able to develop draft professional standards. At the same time, educational organizations of vocational education, along with other interested organizations, are participating in this work.

It should be noted that Russian universities are not able to make sufficient profit to allow the equipping of their laboratories with the most modern technologies. Nevertheless, enterprises and organizations that are interested in finding employment for university graduates are actively involved in the educational process, especially with senior courses.

It is known that professional enterprises and institutions usually have more advanced equipment and technological base when compared to what is available in higher educational institutions. In order to eliminate this gap, basic chairs (university departments) are organized in enterprises. University teachers and representatives from enterprise are working in these foundational positions. The purpose of creating these basic chairs is to effectively convey to bachelors and undergraduates the knowledge and skills necessary for practical professional activity (Listkov & Belousova 2018).

Modern engineers and specialists need detailed, in-depth practical knowledge. That is why, in recent years, higher education institutions have carried out many events and activities. Such events include the opening of government supported technology parks and quantoria (specialized scientific-research communities with students).

In these quantoria, students and undergraduates use the most advanced technologies for learning, including additive technologies such as 3D prototyping, virtual reality, augmented reality, electronics, programming, robotics, and neurotechnologies. The network of such quantoria is quite large and students can choose suitable programs for themselves.

Regarding the direct education of students, the following should be noted. Specialized classes are offered from among existing disciplines. For example, the Faculty of Information Technology at the Russian State Social University offers training programs in the fields of informatics and computer science. These training disciplines include special sections on programming; software and hardware; automated systems; and mobile programming. Other programs are also provided with specialized training modules. These modules include practical programming (the programming of real devices) based on some modern well-known platforms, such as Arduino, Iskra JS, Raspberry Pi, and others. Throughout these courses, students learn about programming environments, software, hardware, and programming languages (e.g. C++/Wiring; JavaScript; Python); and acquire skills for building simple devices using sensors (for sensing light, movement, distance, sound, acceleration, etc.). These devices function according to their programming (movement; signaling with sound or light; information exchange between devices).

At the same time, freshmen students of the aircraft control systems specialty of MAI university (where a co-author of this article works) are taken through summer training sessions, which are aimed at practising entry-level tasks in the field of measurement and data processing, control parameters, and device automation.

The course of study involves guided learning by a teacher, as well as self-directed learning. During self-study, students read textbooks, select projects, study materials, design information processing algorithms, write programming code, and test their performance in virtual development tools.

The study course on software and hardware applications contains practical exercises and tasks, starting at the level of flashing LEDs and progressively increasing in difficulty (Karvinen, Karvinen, & Valtokari 2016; Amperka 2019; LDM-Systems 2019):

- Students use Morse code or other codes to write their own functions.

- Making alarms sound (both simply and through using Morse code or another code).
- Using sensors (tilt sensors, temperature, humidity, light, smoke, ultraviolet radiation, buttons, potentiometers, joysticks, etc.) Students also use specialized sensors: accelerometers, barometers, gyroscopes, magnetometers, and several others, including combined ones.
- Using various indicators: single and multi-color LEDs; LED bars; seven-segment indicators (including multi-digit assemblies of seven-segment indicators); LCD-indicators (including color graphics touch screens); LED matrices; piezo-emitters of sound; and vibro-engines, etc.
- Using motors and servo drives.
- Studying motion control algorithms.
- Studying wireless communications, as well as the connection of automated devices to the Internet.
- The assembly of various devices (automatic and automated), including applications for people with disabilities.
- The assembly of mobile robots.

Further training in this area is carried out by using single-board computers, such as the Raspberry Pi, and students learn the basics of Python along with the other programming languages (Karvinen, Karvinen, & Valtokari 2016; Amperka 2019).

It should be noted that the program provides for the study of domestic analogs of elemental electronics and hardware bases, for example, those of the LDM-Systems company (LDM-Systems 2019).

Students gain knowledge and skills in:

- The development of the Arduino IDE interface, the Espruino IDE, and other programming environments.
- Understanding the basic concepts of a program (variables, cycles, arrays, procedures, functions, conditional statements, etc.).
- Sequential and parallel execution of commands; changes in the modes of object operation and events.
- Usage of libraries (both open source and independent development).
- Usage of widespread electronic components for the creation of electronic circuits for automation and robotics, including domestic production.

- Wireless communication between devices and communication with the Internet.

Scientific research groups for students (Alexandrov et al. 2016) include the following disciplines in their curricula:

1) Student awareness of goals and objectives of the scientific research group (for students enrolled in this group), joint scheduling of work, and discussion and approval of student research topics.
2) Teaching methods of scientific research work (SRW).
3) Conducting SRW on pre-approved topics.
4) Testing the results of SRW at conferences, exhibitions, and competitions; and the publication of research materials.

It is noteworthy that students enrolled in scientific research groups can act as teachers for particular topics (if they have above average knowledge). Hence, the students can not just improve their knowledge, but also attract peers to join the group.

In our experience, in the scientific work circle it takes a few years (two to three) after the start of studying a particular problem for innovative and patentable solutions to come to fruition. There are several reasons for this. Firstly, the technical infrastructure (laboratory, or group, technical circle) may take a while to be organized (at least two years, unless there is a pressing need to hurry).

Secondly, students and undergraduates, although they study for a rather short period of time of only two to four years, become part of a scientific body of researchers and transfer knowledge and skills to students of junior courses. As such, a "virtual chain of knowledge transfer" is formed. Knowledge, projects, and a focus on relevant scientific activity does not occur immediately, but takes at least 2-3 years.

Before designing more serious projects (for example, using elements from the fields of the Internet of things and Industry 4.0), students study the constituent parts and components of future projects, including:

- computing microcontroller devices in the Internet, with cloud resources;
- user-friendly devices with voice and/or sound communication capabilities and gesture control, which are convenient for people with disabilities, and other capabilities;
- browser games;

- elements of the "smart home", including its interactive elements;
- development of diaries as memoranda of important events;
- devices using modern instant messaging, etc.

The training program includes the development of familiarity with simulation programs, such as MATLAB and others. In addition, students use cloud modeling systems.

A necessary part of this training is the preparation of technical documentation that includes: presentation of the technical task; a technical description and instruction manual; and adjustment instructions. These documents are developed in accordance with relevant national standards.

Let us discuss exactly what competencies students receive and how their educational trajectory is constructed.

In accordance with the Federal State Educational Standard 09.03.01 "Computer Science and Computer Engineering" (№ 929 from 09.19.2017), the student must be:

- able to install software and hardware for information and automated systems (OPK-5);
- able to participate in the configuration and adjustment of software and hardware systems (OPK-7);
- able to develop algorithms and programs suitable for practical use (OPK-8).

These abilities are treated as synaptic weights on an artificial neural network, as discussed in the previous section (see Fig. 5) and their level is determined through expert evaluation. The inputs of artificial neural network are micromodules of the disciplines taught. A table was built for this case (see Table 1).

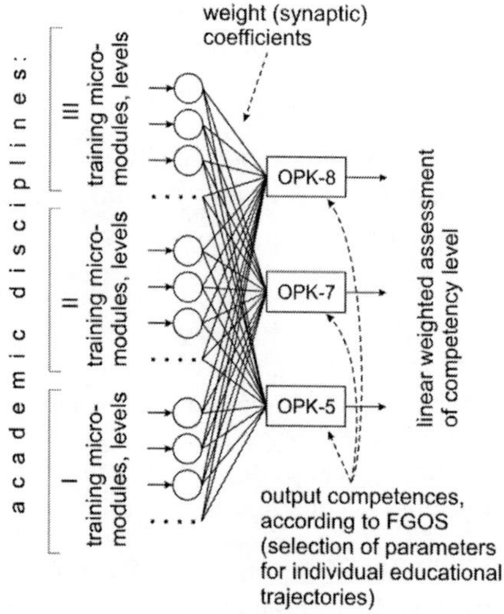

Figure 5. Artificial neural network of competencies

Table 1. Weights of the discipline micromodules in relation to output competencies

Name of discipline	Name of micromodule	Weight for OPK-5	Weight for OPK-7	Weight for OPK-8
Special sections of programming	Write own functions. Level 1	0,10	0,15	0,4
-- “” --	Using sensors. Level 1	0,35	0,2	0,4
-- “” --	Using various types of indicators. Level 1	0,2	0,6	0,5
-- “” --	… (8 micromodules total)	…	…	…
Software and hardware of computer equipment and automated systems	Sensors. Level 2	0,45	0,65	0,5

Name of discipline	Name of micromodule	Weight for OPK-5	Weight for OPK-7	Weight for OPK-8
-- "" --	Motors and servo drives. Level 1	0,2	0,2	0,2
-- "" --	Motion control algorithms. Level 1	0,7	0,5	0,55
-- "" --	Wireless communications. Level 1	0,3	0,65	0,35
-- "" --	... (16 micromodules total)	...	...	...
...	...	...	...	...
Scientific research work group for students (SRW)	User-friendly devices. Level 4	0,9	0,9	0,9
-- "" --	IoT devices and projects. Level 4	0,9	0,9	0,9
...	...	...	...	...
	Result (weighted value):	max 1 (SRW gives additional points)	max 1 (SRW gives additional points)	max 1 (SRW gives additional points)

The results show that some change in the learning path is quite possible (expressed in the variability of the choice of micromodules), while successfully achieving the final result.

The individual areas in which students were trained in accordance with the specified methods are presented below.

Artificial Intelligence

Training of students in the field of artificial intelligence (AI) traditionally begins with the study of artificial neural networks. These can successfully help resolve practical problems of pattern recognition. The Hamming-Hopfield neural network emulation project was carried out (Zinoviev & Zinovieva 2013; Balashova, Zinoviev, & Zinovieva 2013). The network consists of two layers. A detailed description of these layers follows. The first and second layers each have *m* neurons, where *m* is the number of samples. The neurons of the first layer have *n* synapses each connected to the network inputs (forming a dummy null layer). The neurons of the second layer are interconnected by inhibitory (negative feedback) synaptic connections. A single synapse with positive feedback for each neuron is connected to its own axon. The idea of the network is to find the Hamming distance from the test image to all samples. The Hamming distance is a number of different bits in two binary vectors. The network must select a sample with a minimum Hamming distance to an unknown input signal, with the result that only one output of the network corresponding to the sample will be activated.

This artificial neural network solves a particular problem with associative memory. A certain set of binary signals (images in this case) are considered to be exemplary. The network should be able to select (given incomplete input, the network is designed to remember all previously collected data) the corresponding sample (if any) or “give a conclusion” that the input data does not correspond to any of the samples from an arbitrary non-ideal signal applied to its input. Since in this case, the signals are arbitrary images, displaying the data from the network output in graphical form indicates whether the result completely coincides with one of the model (if successful) or the “guessing” of the network in case of failure.

An experiment on the compilation of graphic object recognition was conducted by the authors over a long period of time. This experiment was performed with a different degree of distortion to the original graphics data, as well as with a different amount of basic configurational elements. As a result, a pattern with the primary deteriorating feature of object recognition and an increasing base was revealed. This problem can be solved theoretically by complicating the neural network.

Fig. 6 shows the interface window of the software. It displays sample images, as well as the results of the experiment.

Figure 6. User interface of the software. The user can see some sample images and buttons with which to conduct experiments on image recognition in relation to the degree of image deconstruction

Other projects in the field of artificial intelligence use artificial neural network algorithms in the development of expert systems (for example, evaluating a university student's rating). University departments often face similar problems in grading the students' performance. This problem involves not only grading according to academic performance, but also according to other indicators, such as: student attitude (attendance, lateness, classes skipped); participation in scientific and public events; appearance; behavior; and other educational and extracurricular activities. A sound analysis of this information is necessary to identify a reasonable distribution of a limited number of awards or special scholarships, etc.

In addition to a comprehensive evaluation (general rating), the "individual ratings" of students are required, for example, "the student is a professional one", "the student is a novice scientist", "the student is a public individual", and "the student is a neat and responsible", etc.

For employers, it is often important to use certain special criteria to evaluate a student, such as creativity, independence in undertaking tasks, problem-solving skills, and special skills, etc.

Existing systems do not fully take into account a student's activities. This is important for solving the following tasks:

1) How to develop a comprehensive and flexible system of criteria for evaluating student activities.

2) How to develop a computer system capable of calculating general and individual ratings of university students according to the developed criteria.

Algorithms of information processing using artificial neural network (forward propagation neural network) can help to solve these problems.

The weights of the synaptic connections between neurons are determined by expert evaluation of each of the rating indicators. Indicators are grouped, and weights are determined for each group (Erpelev, Simonov, & Rubankova 2019).

There is another direction of design in the field of artificial intelligence working with the KUKA palletizing robot. This robot was recently installed in the Technopark of the Russian State Social University. Development of software and hardware controls for special hooks and additional gadgets, both stationary (in the capture range) and attached to the moving parts of the palletizing robot, are planned. One of these projects is focused on working with handicapped people (see Fig. 7).

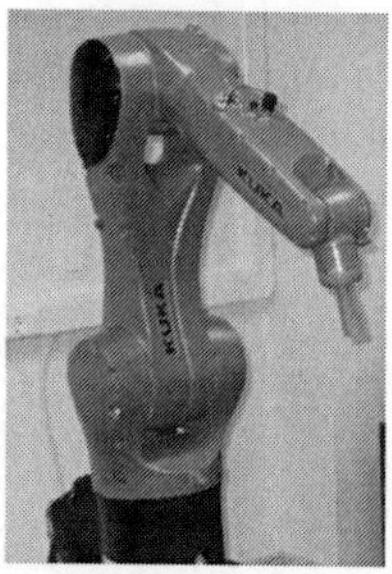

Figure 7. KUKA palletizing robot, intended for various projects through software and hardware controls and executive mechanisms using Arduino, Raspberry Pi, and Iskra JS. Projects are aimed at working with handicapped people as well

It should be noted here that the micromodules in this case are arranged to take into account several factors: labor intensity; degree of compliance with competencies; level of complexity of implementation (Level 1, Level 2… etc.), i.e. the presence of those or other input competencies.

Projects for Automation and Robotics

Student projects have been carried out on various topics, including “Smart home”; “Room parameter evaluation system with LED color display”;

"Weather station project"; "Human psycho-physical condition diagnostic device"; "Device-based on the palletizing robot KUKA for handicapped people"; and "Intellectual self-learning control system for movement of an object based on neural network technologies"; as well as the development of various other projects using microcontroller boards. All these projects are united by the common themes of robotics, automation, and artificial intelligence.

Fig. 8 shows the operation of the smart home (Bakin, Venina, Derbysheva, Erpelev, Ovsyannikov, Petrova, Rubankova & Simonov 2018; Katorgin 2019). The general scheme is presented here. It consists of two parts functioning in automatic mode or under the control of external commands. The device was made using several microcontroller boards: Arduino and Iskra JS. These boards work together (see Fig. 9).

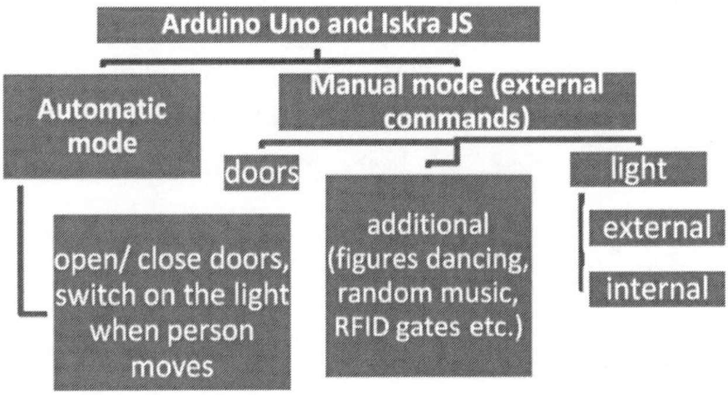

Figure 8. Smart home operation scheme. The two branches represent the automatic function mode and operation under the control of external commands

Figure 9. Smart home models assembled with various options and different functionalities

Students developed a system for evaluating room parameters with an LED color display (weather station) using the studied microcontroller systems, see Fig. 10 (Katorgin 2019).

Figure 10. System for room parameter evaluation with LED color display (weather station). The indicators on the right of the front panel are adjusted in accordance with relevant sanitary and epidemiological standards

All student projects use freely accessible software.

The following projects were developed by students involved in a research group. The students designed a diagnostic device of the psycho-physical state of a person on the basis of photoplethysmography and cardiography, see Fig. 6 above (Pulsometer. Lesson #27 2019). Results were obtained and reported at a conference (Pikova, Simonov, & Boyko 2019), see Fig. 11.

Currently, a number of promising projects are being implemented, for example, palm thermography diagnostics of personal characteristic properties (Nidilko & Mitarova 2019).

One of the most engaging projects is the design of a system for the optoelectronic measurement of vibrations and displacements, using the individual image of an object (Simonov 1993; Simonov & Kuzin 2003). Raster conjugation was used in this particular project. Specifically, the project involved overlaying the optical image of the measured object (or a contrasting light mark on it) on the analyzer. Dependencies of the measured result on a number of parameters are formed. Analysis of these parameters reveals, that some are informative, while others are uninformative.

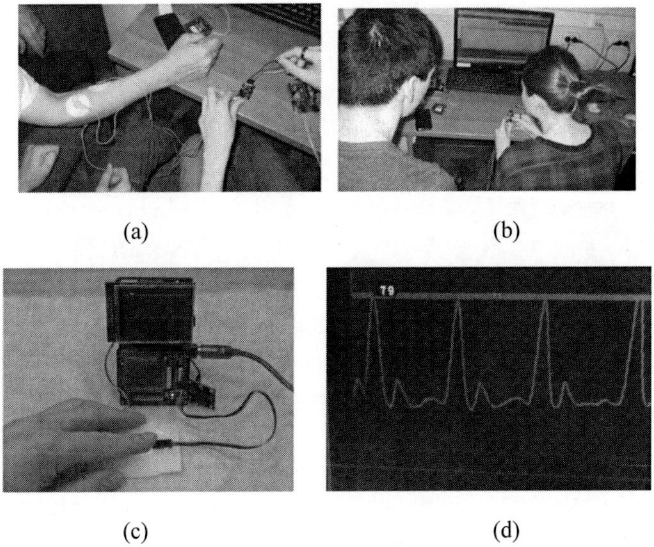

Figure 11. Device for diagnosis of the psycho-physical state of a person on the basis of photoplethysmography and cardiography: (a) sensor test; (b) study of tabular data—results of measurement; (c) photoplethysmography sensor (pulsometer. Lesson #27 2019), rewritten software (one of the specialized libraries); (d) type of photoplethysmogram

The formula for mathematical modeling is introduced below. Let the light mark, which is projected onto the photocathode, be described with a two-dimensional function of brightness E(x,y) of approximately symmetrical shape. Sensing of light information is conducted using an aperture H(x,y). Vibration measurements along one axis (X) produce a voltage at the output of the light-electronic transformer:

$$U_{out}(t) = K_1(x)\, R_l \int \sigma\, E(x(t))\, H(x(t))\, dx \tag{1}$$

where $K_1(x)$ is the function of non-equal sensitivity of the aperture along the X axis; R_l is loading resistor; E(x(t)) and H(x(t)) are the corresponding light and aperture functions; and σ is the work area. For the selected principle of measurement, the displacement of the object, *xob*, produces an output voltage (if the scanned function is periodical):

$$U_{out}(t) = K R_l \int \sigma E (x - x_{ob}) H(x - F(t))\, dx \tag{2}$$

($K_1(x) = K = const.$)

For every special kind of function, E(t), H(t), and the scanning function F(t), we can derive Uout(t). Let E(t) and H(t) be indicator functions (as with E(t) = c inside the interval [-e, e], which is equal to zero outside the interval; H(t) = b inside the interval [-d, d], which is equal to zero outside of it); and F(t) = Acosωt, where ω is the scanning frequency. Then, if xob = const Uout(t) it can be presented as a Fourier series (the expression is rather large and will not be presented here).

As a result, the amplitude of the first harmonic (k = 1) is:

$$a_1 = b\, c\, (2/\pi)\beta(\sin\varphi_3 - \sin\varphi_0) - x_{o6}(\sin\varphi_1 + \sin\varphi_2 - \sin\varphi_0 - \sin\varphi_3) - \alpha(\sin\varphi_2 - \sin\varphi_1) + ½\, A(\varphi_3 + \varphi_0 - \varphi_2 - \varphi_1) + ½\, (\sin 2\varphi_3 + \sin 2\varphi_0 - \sin 2\varphi_2 - \sin 2\varphi_1))) \tag{3}$$

where:

$$a = e - d,\ \beta = e + d,$$
$$\varphi_0 = arc\cos(\min(1, (\beta - x_{ob})/A)),$$
$$\varphi_1 = arc\cos(\min(1, (\alpha - x_{ob})/A));$$
$$\varphi_2 = arc\cos(\max(-1, (-\alpha - x_{o6})/A)),$$
$$\varphi_3 = arc\cos(\max(-1, (-\beta - x_{o6})/A)). \tag{4}$$

Equation (3) is the equation of measurement. As such, a scheme was developed to compensate the measurement results (see Fig. 12). At this stage, this scheme is tested using a microcontroller base.

In accordance with the principles described above, a model of a vibration/displacement measuring system was constructed. Several kinds of objects were tested, for which it is difficult to use traditional vibration testing systems. These objects are as follows: fluctuations (vibro-displacement) of human skin areas—projections of the heart muscle onto the surface of the skin in the thoracic region (studies were conducted in accordance with the recommendations of a specialized medical institution); aviation engine blades (test for vibration resistance); aviation turbine coolers; elements of printed circuit boards in electronic devices; parts of various constructions; acoustic loudspeaker diffusers; active objects (light emission); several parts of engines in the automobile industry; distance measurements (hundreds of meters); and some other applications. The above model was also used for rotor balancing.

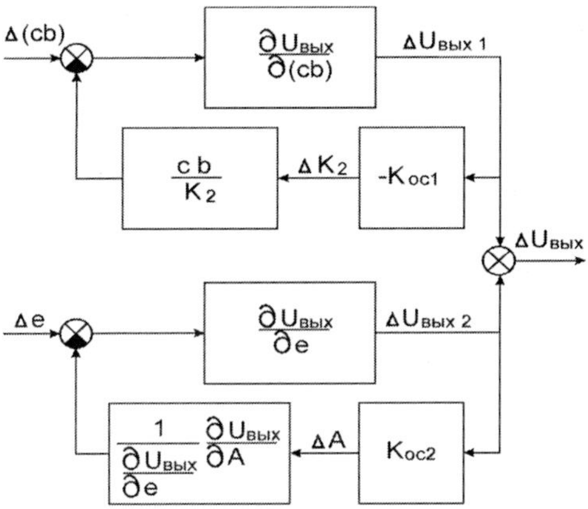

$$\Delta K_2 = -K_{oc1}\,\Delta U_{вых1}$$
$$\Delta A = K_{oc2}\,\Delta U_{вых2}$$
$$\Delta U_{вых} = \Delta U_{вых1} + \Delta U_{вых2}$$

Figure 12. Introducing compensation into the measurement results

Experiments related to the use of elements in the field of artificial intelligence were conducted. The main conclusion of this research was that objects, subjected to vibration testing, have rather complicated structures. For example, printed circuit boards have a large number of diverse elements; as do aviation electronics and other distinctive types of devices. A vibration test was performed on these devices in order to improve their construction and to develop stability against vibration. It is obvious that the vibration behavior of such boards needs to be thoroughly tested. As such, a measuring system should be able to acquire information on the vibration of a large number of areas ("objects") of a board, store it, and then display it in an appropriate way. It is to be noted that the measuring system is supposed to deal with various kinds of objects, varying in size, shape, optical reflection ability, and distance to it, etc. In order to solve this problem, an intellectual robotic system was designed. This system is of an adaptive type and its functions are as follows. In general, the vibration measuring of a robotic system (VMRS) should recognize the borders of an object and be able to deal with the spots, lines, and curves of the object, depending on the user's goals. The VMRS should automatically scan an area defined by the user, form a field of the object's vibration, and then display it on a computer screen or make a hard-copy.

In particular, a VMRS should be able to measure vibrations in objects of different types: spots on object surfaces; straight lines; and simple and complicated curves. While performing the scanning procedure, it was concluded that the VMRS tracks a line or curve, according to user choice. In the case of studying points, the VMRS searches for the most relevant points and measure their vibrations one after another. Adaptive algorithms for the control of a system's mode were implemented in this research. The area to be measured is chosen by the operator.

It is also notable that VMRS operations are conducted in two modes: active (using the special lighting system) or passive (without it). During operation, even modern kinds of sensor have a limited speed of frame change, which would not allow measurements to be taken at high frequencies (up to hundreds of kHz).

Several parts of the vibration measuring robot-system were developed and tested. As a result, existing errors were eliminated during the measuring of the object vibrations. The optical reflective ability of these objects varied by up to 100 times; the shape varied about 30 times (in passive mode); and the lines of various features were tracked. This research used a semi-automatic algorithm for the adaptation of contrasting objects during the scanning procedure. Consequently, the number of main features of various objects (along some geometrical parameters) were determined, defined, and calculated. In order to construct these distinctive types of objects, the measurements were classified and used to define the main types of "light functions" and their major features (Simonov & Kuzin 2003).

The above mentioned features were classified in order to prepare for the task of creating a database. Separate and simultaneous measurements of both linear and angular vibrations were undertaken.

Projects for Various Purposes

Projects have focused on recreation and games with electronic devices for assessing a person's in-game reactions.

One of the devices developed has a built-in sound signal, two buttons, a display device, and can present results directly. The sound signal is generated at a random point in time; after it sounds, the participant must press the button five times as quickly as possible. After each attempt, the screen displays the results for each participant, indicating how many times they pressed the button faster than their opponent. This allowed the reactions of the two participants to be compared. This project used a game in the form of a competition. It is used to determine the state of a person.

Another device was developed to determine the response of a person. It consists of actuators (speakers) that mimic a sharp blow. The participant moves the palm of one hand, while pressing the button with the other hand as quickly as possible. After performing a series of experiments with one hand, the hands are changed, and the series of experiments is repeated.

These devices are used as part of the installation presented above to assess the psycho-physiological state of a person (see also Fig. 6 above).

As part of continuing research into the psycho-physiological state of a person. This was as a result of interaction with the State Public Institution of the City of Moscow Center for Sports Innovative Technologies and National Teams of the Department of Physical Culture and Sports of Moscow (GKU CSTiSK Moskomsport) and analysis of marksmanship was performed.

The first of these developments used a laser shooting device to study the biathlon discipline. In this game, each participant was given three attempts to shoot (five shots each time) in intervals between jogging (or completing) a certain distance. The scoring software was configured in such a way that it is possible to shoot accurately, but if the distance is covered too slowly then the participant will receive few points. It is possible to cover the distance quickly, but this will reduce shooting accuracy due to physical over-stimulation and respiratory stress. In such a case, the participant will also receive few points. As such, it is important for participants to choose the optimal shooting mode for themselves, as well as completing the distance. This device has been tested on more than 50 students and adult employees of enterprises.

The second one was a device for measuring and recording oscillations of the hand or the barrel of an athlete's gun. The device comes in two versions.

The first is completely autonomous and similar to a pistol. The device contains an oscillation sensor and a recording component. This device summarizes the deviations of the athlete's hand from the central position (for example, the axis of aiming) over short time intervals (5 seconds). After the specified period of time, the readings are indicated. This device was developed to provide an inexpensive mobile device for the preliminary assessment of the readiness of a competitive shooter (how static is the stand of the shooter and how large are the fluctuations of their hand when aiming).

The second is a similar device. However, this option has a remote sensor that is attached to the barrel of the weapon (its fixed part) or to the hand of an athlete. The device functions in the same way. This device achieves rapid assessment of the oscillations of the gunbarrel or the hand of an athlete while shooting.

It should be noted that these two devices can be used as part of a more general complex assessment of the human psycho-physiological state (see Fig. 6). This is because the psycho-physiological state of a person is due to their motor activity.

It is known that in the operation of any system, an assessment should be produced according to the relevant criteria of a set of parameters: input, output, and intermediate points.

At the same time, a mandatory evaluation of the system needs to be carried out as part of other higher level systems, in accordance with predetermined criteria. Here the goal is to determine how the system in question meets the requirements of other systems.

In recent years, in the Russian system of higher education of the use of grade-points is widespread, as a convenient way of developing both binary grades of the good/bad type and a grade scale on how good or bad a student's work is. Such assessments have been applied to many aspects of university assessment, including the work of teachers and student activities.

Such estimation procedures can have a positive impact. For example, with respect to teaching staff, these procedures can reveal a general picture about teaching staff's activities in a department and show who is lagging behind in publishing or other types of activity, and who is engaging in public facing events, etc. At the same time, a university is not a production process in a direct sense. It is necessary to apply estimation point systems with great caution, due to the essence of the learning process being a creative process. As such, other teachers may not achieve the required number of publications and be awarded a low number of points, but at the same time they may be busy producing innovative and patentable solutions to problems.

Requirements for universities in the form of Education Standards FGOS (currently FGOS3++) are being developed. Competencies presented in FGOS3++ and other corresponding documents need to be filled out with specific content. These competencies are to be achieved by universities at their own discretion, depending on the specialization, means, and capabilities of the teaching staff.

Depending on the requirements of the industry and what was studied, a graduate may turn out to be in demand or not (see Artificial Intelligence).

The correspondence of the concrete content of teaching courses to the requirements of FGOS3++ can also be determined using neural network representations (Sviridov 2001). The requirements of FGOS3++ and the corresponding documents can be presented in the form of synapse settings of the first artificial neural network layer. The specific content of these courses is the input signal vector (Order of the Ministry of Education and Science of the Russian Federation No. 929 of 09.19.2017). As a result, an assessment of compliance with the requirements of the standard can be undertaken (Titov, Simonov, & Erpelev 2019). Detailed materials on this development are presented in Sviridov (2001).

It is clear that the above problems and trends are part of a global development. For example, Siau (2017) discusses the impact of artificial intelligence, robotics, and automation on higher education. In this article, the author examines in detail the relevant trends in higher education related to the development of AI, robotics, and automation. Siau notes that: “in a short term (i.e., the next 5-10 years), higher education will still need to train students for industry” (2017). But at the same time higher education itself will change because “what can be automated will be automated” (Siau 2017).

Several ideas for starting projects can be found in books by Norris (2017) and Grimmett (2014).

Some world-class universities offer specialized training programs on this subject. Candidates for training at Tampere University of Applied Sciences are, to some extent, are required to be familiar with the basics of programming, digital logic, and processors, as well as to have knowledge of machine learning and control (degree program 2019: Information Technology, Robotics and Artificial Intelligence). The AI program at the

University of Michigan comprises a multidisciplinary group of researchers conducting theoretical, experimental, and applied investigations of intelligent systems (program: Artificial Intelligence (AI). 2019).

Students are often interested in the influence of AI, robotics, and automation on their workplaces (currently and in the future) and whether intelligent algorithms and robotic production will lead to mass unemployment. Such problems are discussed in Wisskirchen et al. (2017). With the constant development of technologies, educational systems and legislators are faced with the task of solving new problems. Companies, lawyers, and society in general also face these problems. Wisskirchen et al. (2017) give several examples of how AI will fundamentally change the world of employment.

It is necessary to note the following. The choice of an individual educational trajectory is similar to the choice of a set of dishes in a buffet. Yes, the visitor is offered some choice, but only within the framework of what was cooked by the chefs, and nothing more. With regard to educational institutions, this means that an educational institution can only teach students what teachers can do. It is no accident that an extensive study (State of the Art Report: "Governance of educational trajectories in Europe" 2015) noted:

> "The quality of education depends on the infrastructure of educational institutions. In this regard, teachers and their training are crucial to the quality of an educational system" (State of the Art Report *Governance of educational trajectories in Europe* 2015 141).

Conclusion

This chapter has discussed the practical aspects of modeling individual educational trajectories of students in the fields of informatics and computer science. Additionally, classes in the fields of programming, electronics, and robotics are considered. The general orientation of these classes require fundamental training in artificial intelligence, robotics, and automation. Modern aspects, such as the Internet of things, the smart home, and some initial aspects of Industry 4.0 are touched upon, for example, in the context of the research group. In doing so, students learn to design and assemble products in the fields of electronics, programming, automation, and robotics. The achievements of artificial intelligence, for example, artificial neural networks of various types, are also used. Design results are discussed in classrooms, published, and presented through exhibitions and conferences. It is important to note that successful students can easily find work in various enterprises, including defence enterprises (and the Ministry of Defence).

A complex project proposed by students and undergraduates used a system for measuring vibration and displacement. These studies described error compensation occurring because of differences in the size and shape of objects, utilizing the fact that objects usually emit or reflect different amounts of light. Without the use of compensation, the total error would be in the range of hundreds of percent. This is why all kinds of measurement errors were analyzed, classified, and mathematically modeled, and special contours were created to compensate for errors.

Acknowledgments

The authors would like to extend their sincere gratitude to the administration of the Department of Information Systems, Networks and Security, the Faculty of Information Technologies of the Russian State Social University for the opportunity to conduct their research. The authors would also like to show their appreciation to all the students who took part in the research and who have presented their results and spoken at national and international conferences. Special appreciation is held for our colleagues from other higher educational institutions—the Moscow Aviation Institute (National Research University) and the Moscow City Pedagogical Institute—for their cooperation and valuable comments made while sharing experiences.

Bibliography

Alexandrov, Pavel, Tatiana Dmitrieva, Marina Malashenko, et al. 2016. "*A promising direction for the development of scientific and technical creativity of young people - a circle of fundamentals of electronics and robotics*" (abstracts of the report), Materials of the XXVII international conference "Modern information technologies in education", June 28. Troitsk Moscow: DOGM, "Baitik" foundation, ANO "Information technology in education ITO. Pp. 454-455.

Bakin, Mikhail, Ekaterina Venina, Anna Derbysheva, Alexey Erpelev, Nikita Ovsyannikov, Elena Petrova, Anastasia Rubankova, and Vladimir Simonov. 2018. "*Application of smart home technologies in boarding schools and other organizations of the social sphere*", Collection of reports of the II All-Russian Forum of Young Scientists "Wealth of Russia". Held in Moscow, December 10–11, 2018), Moscow State Technical University named after N.E. Bauman

(National Research University)". Moscow: Publishing House MSTU. N.E. Bauman. Pp. 156-157.

Balashova, Natalia, Vladimir Zinoviev, and Nadejda Zinovieva. 2013. "*The use of neural network algorithms in the development of simple expert systems (for example, assessing a university student's rating*)", a project prepared for exhibition NTTM-2013. Moscow: Department of ISIT IMMI Moscow City Pedagogical University. Unpublished.

Degree programme 2019: "*Information Technology, Robotics and Artificial Intelligence*". Tampere University, Tampere University of Applied Sciences, Finland.

https://www.tuni.fi/en/study-with-us/information-technology-robotics-and-artificial-intelligence.

Erpelev, Alexey, Vladimir Simonov, and Anastasia Rubankova. 2019. "*The use of free software in the educational process on the example of the development of Smart Home devices using Arduino and Iskra JS microcontrollers*". Fourteenth Conference "Free Software in Higher School": Conference materials. Pereslavl: ANO "National Supercomputer Forum", Software Institute Systems RAS, Institute of Logic, Basalt SPO. January 25–27, 2019. Resp. Editor Cherny V.L. Moscow: MAKS Press. Pp. 34-36 https://www.basealt.ru/fileadmin/user_upload/Programma_14_final-5.pdf.

Federal State Educational Standard of Higher Education. Bachelor's level. *Training direction 09.03.01. Informatics and computer engineering*. Approved 01/12/2016. http://fgosvo.ru/uploadfiles/fgosvob/090301.pdf.

Grimmett, Richard. 2014. "*Raspberry Pi Robotic Projects*". Birmingham B3 2PB: Packt Publishing Ltd. UK.

Ibatullin, Rinat R. and Ellina S. Anisimova. 2016. *Construction of individual educational trajectory of students based on e-learning*. Publisher: IEEE / Proceedings of the 2016 IEEE 10th International Conference on Application of Information and Communication Technologies (AICT) https://ieeexplore.ieee.org/document/7991799.

Karvinen, Tero, Kimmo Karvinen, and Ville Valtokari. 2016. *Making Sensors. Projects of touch devices based on Arduino and Raspberry Pi.* Moscow: ID Williams Ltd.

Katorgin, Maksim. 2019. "*Use of free software in a student project of a meteorological station based on the Arduino MEGA board*". Fourteenth Conference "Free Software in Higher School": Conference materials. Pereslavl: ANO "National Supercomputer Forum", Software Institute Systems RAS, Institute of Logic, Basalt SPO. January 25–27, 2019. Resp. Editor Cherny V.L. Moscow: MAKS Press. Pp. 101-103. https://www.basealt.ru/fileadmin/user_upload/Programma_14_final-5.pdf.

Listkov, Vjacheslav, and Ella Belousova. 2018. *Prospects for the development of educational technologies*. In the collection of Technologies in Education. 2018: Materials of the International Scientific and Methodological Conference. April 23–30, 2018/PEI HE Tsentrosoiuza RF "SibUPK". Novosibirsk, 290 p., pp. 21-27.

Nidilko, Marina, and Djamilia Mitarova. 2019. "*Development of a project on a microcontroller board*". Fourteenth Conference "Free Software in Higher School": Conference materials. Pereslavl: ANO "National Supercomputer Forum", Software Institute Systems RAS, Institute of Logic, Basalt SPO. January 25–27, 2019. Resp. Editor Cherny V.L. Moscow: MAKS Press pp. 36-38.
https://www.basealt.ru/fileadmin/user_upload/Programma_14_final-5.pdf.

Norris, Donald J. 2017. "*Beginning Artificial Intelligence with the Raspberry Pi*". New York, Barrington, New Hampshire: Springer Science+Business Media, Apress Media, USA. https://ru.scribd.com/document/361666126/Beginning-Artificial-Intelligence-With-the-Raspberry-Pi#.

Order of the Ministry of Education and Science of the Russian Federation No. 929 of 09/19/2017 "On approval of the federal state educational standard of higher education, bachelor degree in the field of training 09.03.01 Informatics and computer engineering".
http://fgosvo.ru/uploadfiles/FGOS%20VO%203++/Bak/090301_B_3_12102017.pdf .

Pikova, Alisa, Vladimir Simonov, and Lyubov Boyko. 2019. "*Using free software for designing devices for diagnosing of psychophysical state of a person*", Fourteenth Conference "Free Software in Higher School": Conference materials. Pereslavl: ANO "National Supercomputer Forum", Software Institute Systems RAS, Institute of Logic, Basalt SPO. January 25–27, 2019. Resp. Editor Cherny V.L. Moscow: MAKS Press pp. 114-117.
https://www.basealt.ru/fileadmin/user_upload/Programma_14_final-5.pdf.

Pozharkova, Irina. N., Noskova Elena E., and Troyak Evgenii Yu. 2018. *Formation of individual educational trajectory as a component of practice-oriented learning environment*. In Theory and Methods of Professional Education. Pp. 179-192.
https://cyberleninka.ru/article/v/formirovanie-individualnoy-obrazovatelnoy-traektorii-kak-komponenta-praktiko-orientirovannoy-sredy-obucheniya.

Program "*The Artificial Intelligence (AI)*". 2019. University of Michigan, USA. https://cse.umich.edu/eecs/research/area.html?r_id=2.

Pulsometer. Lesson #27 on the Website of the company "iArduino". 2019. https://lesson.iarduino.ru/page/urok-27-pulsometr/.

Recommender systems in online education. Adaptive learning. 2017. https://m.habr.com/ru/company/stepic/blog/325206/. 29 March.

Schaffhauser, Dian. 2018. *The Next Frontier of Adaptive Learning.* 14 November. https://campustechnology.com/articles/2018/11/14/the-next-frontier-of-adaptive-learning.aspx?fbclid=IwAR3Jlrd7TdonPoVeuC4gk38Ykfp OvkNa6WFvRKHvTxUrLLiNkNug7rBbPWw&m=1

Siau, Keng. 2017. "*Impact of Artificial Intelligence, Robotics, and Automation on Higher Education*". Twenty-third Americas Conference on Information Systems, Boston: Missouri University of Science and Technology.

Simonov, Vladimir, and Alexander Kuzin. 2003. "*Intelligent system for optoelectronic surface vibration testing*", Report on the conference ICIASF'03. Proceedings of the 20-th International Congress on Instrumentation in

Aerospace Simulation Facilities. Deutsches Zentrum fur Luft- und Raumfahrt e.V. (DLR), Germany, Goettingen: IEEE Catalog Number 03CH37501. No 4.3. August 2003 pp.115-117.

Simonov, Vladimir. 1993. *Optical vibration measuring system which uses the image of the object* / Report on the conference ICIASF'93. Proceedings of the 15-th International Congress on Instrumentation in Aerospace Simulation Facilities. French-German Research Institute, France, Saint Louis: IEEE Publication 93CH3199-7. September 1993. Pp.30.1-30.5.

State of the Art Report "*Governance of educational trajectories in Europe*". Working paper 07. Deliverable Nr. 6. University of Frankfurt/University of Tübingen, University of Ljubljana, EHESP, Rennes. March 2015.

Sviridov, Alexander. 2001. *Neural networks and neurocomputers*. Moscow: Publishing house of Moscow State Social University "Soyuz".

Titov, Evgeniy, Vladimir Simonov, and Alexey Erpelev. 2019. *Classes on Artificial Intelligence, Robotics and Automation with Students of Technical Specialties* / Proceedings of the 24th Conference of Open Innovations Association FRUCT (Finnish-Russian University Cooperation in Telecommunications). Moscow, Russia, 8-12 April. Moscow Technical University of Communications and Informatics, April 11. Pp. 760-767. Publication IEEE Xplore.
https://fruct.org/publications/abstract24/files/Tit.pdf
https://ieeexplore.ieee.org/stamp/stamp.jsp?tp=&arnumber=8711952.

Website dedicated to the development of prototypes and products in the field of electronics, programming and robotics. Company "Amperka". 2019.
http://amperka.ru/.

Website of the company "LDM-Systems". 2019. https://ldm-systems.ru/.

Wisskirchen, Gerlind, Blandine Thibault Biacabe, Ulrich Bormann, Annemarie Muntz, Gunda Niehaus, Guillermo Jiménez Soler, and Beatrice von Brauchitsch. 2017. "*Artificial Intelligence and Robotics and Their Impact on the Workplace*". IBA Global Employment Institute, April.
https://www.ibanet.org/Document/Default.aspx?DocumentUid=c06aa1a3-d355-4866-beda-9a3a8779ba6e.

Zinoviev, Vladimir, and Nadejda Zinovieva. 2013. "*Application of neural network technologies in artificial intelligence and robotics*", performance at the competition "Student Science 2013". Moscow: Department of ISIT IMMI Moscow City Pedagogical University. Unpublished.

CHAPTER 3

INNOVATIVE APPLICATIONS AND SERVICES

Editor: Alexey Kashevnik

Innovative applications play a significant role in modern life. Robotics, value added services, and intelligent mobile applications are widespread in people's daily lives and the functionality of the modern smartphones goes far beyond making calls. Smartphone devices allow the implementation of calculations and proactively provide a user with multi-modal information, as well as connecting him/her to a virtual world in any physical location. Modern information and communication technologies are widely used by the research community. This chapter presents some papers related to the development of research methodologies, as well as prototype implementation in the development of innovative applications.

CHAPTER 3.1

TOWARDS BLOCKCHAIN-BASED MULTI-AGENT ROBOTIC SYSTEMS: ANALYSIS, CLASSIFICATION AND APPLICATIONS

ILYA AFANASYEV[1], ALEXANDER KOLOTOV[1], RUSLAN REZIN[1], KONSTANTIN DANILOV[1], MANUEL MAZZARA[1], SUBHAM CHAKRABORTY[1], ALEXEY KASHEVNIK[2,3], ANDREY CHECHULIN[2,3], ALEKSANDR KAPITONOV[3], VLADIMIR JOTSOV[4], ANDON TOPALOV[5], NIKOLA SHAKEV[5], SEVIL AHMED[5]

1 - Innopolis University, Innopolis, Russia
i.afanasyev@innopolis.ru, a.kolotov@innopolis.ru, r.rezin@innopolis.ru, m.mazzara@innopolis.ru, s.chakraborty@innopolis.ru
2 - SPIIRAS, St. Petersburg, Russia
3 - ITMO University, St. Petersburg, Russia
alexey@iias.spb.su, chechulin@comsec.spb.ru, kapitonov.aleksandr@itmo.ru
4 - University of Library Studies and Information, Sofia, Bulgaria
v.jotsov@unibit.bg
5 - Technical University of Sofia, Branch Plovdiv, Bulgaria
topalov@tu-plovdiv.bg, shakev@tu-plovdiv.bg, sevil.ahmed@tu-plovdiv.bg

Abstract. Decentralization, immutability, and transparency are characteristics that make blockchain technology one of the most innovative technologies of recent years. This paper presents an overview of solutions for multi-agent robotic systems based on blockchain technology and an analysis and classification of this emerging field. The reasons for implementing blockchains in a multi-robot network include increasing the interactive efficiency between agents by providing a more trusted information exchange; reaching a consensus in “trustless” conditions; assessing robot productivity; detecting performance problems; identifying intruders; allocating plans and tasks; deploying distributed solutions; and facilitating joint missions. Blockchain-based applications are discussed so as to demonstrate how a distributed ledger can be used to extend the quantity of research platforms and libraries for multi-agent robotic systems.

1. Introduction

The successes demonstrated in recent years in the integration of robotic systems, wireless sensor networks (WSN), cloud computing, distributed planning and management, and distributed ledgers supports an optimistic outlook towards increasingly popular technological solutions, such as the Internet of robotic things (IoRT) (Ray, 2016; Simoens et al., 2018; Batth et al., 2018; Mahieu et al., 2019; Afanasyev & Mazzara et al., 2019) and blockchain-based multi-agent robotic systems (MARS) (Kapitonov et al., 2017; Ferrer, 2018; Strobel et al., 2018; Afanasyev & Kolotov et al., 2019). It is known that one significant problem in developing multi-robot systems is related to the design of strategies for their coordination in such a way that the robots can effectively perform their operations and reasonably coordinate the allocation of tasks amongst themselves (Basegio et al., 2017). Real-world scenarios usually require the use of heterogeneous robots and the performance of tasks of various structures, constraints, and complexity. The method of task distribution for decentralized solutions is appropriate, since the use of autonomous multi-robot systems in complex scenarios becomes limited and inefficient, and centralized solutions pose the danger of failure of the entire system. Since system agents have to share a large amount of information, the requirements for ensuring quality of communication in decentralized systems are increasing; these include such important functions as maintaining data integrity, resilience, and security in accessing data. As such, the implementation of blockchain technology for the interaction and coordination of multi-agent robotic systems offers a reasonable solution to dynamic and decentralized task distribution.

Studies of distributed ledgers demonstrate that decentralization and immutable record technologies make blockchain one of the most powerful innovations: in a decentralized network, intercepting the majority of network nodes in a cyber-attack appears economically impractical[1]. However, many popular blockchain solutions suffer from issues such as scalability, delay, and low throughput[14]. For example, Bitcoin can process less than 10 transactions per second (TPS), while Ethereum can process up to 40 TPS. This is clearly not enough when compared to the 2,000 daily TPS of VISA (which can theoretically increase up to ~50,000 TPS). Since the blockchain network grows as both users and transactions increase, verification

[1] "Blockchain faq", https://medium.com/edchain. Accessed: 2019-07-07

of transactions slows down the transaction process and throughput. This is known as the classical Blockchain Trilemma, i.e. of the desirable characteristics of decentralization, scalability and security, only two out of the three are achievable at any one time[2]. One scaling method that does not compromise security or decentralization is called "sharding". This involves fragmentation of the available dataset into smaller datasets called "shards"[14,15]. Although scalability and speed are not as critical to multi-agent robotic systems (MARS) as they are to financial and big data-based systems, they are nevertheless very sensitive to delays and throughput of information channels during data exchange between agents.

The literature presents many algorithms for distributed consensus, each of which has its own distinctive features, advantages, and disadvantages. A summary of the most important distributed consensus algorithms, including Proof of Work (PoW), Proof of Stake (PoS), Proof of Activity (PoAc), Proof of Burn (PoB), and Proof of Capacity (PoC), etc., are presented in the review by Andoni et al. (2019). The methodologies used to achieve consensus in blockchain networks largely determine their key performance characteristics, including scalability, speed of transaction, completeness of transaction, security, and resource consumption. Each method requires a procedure to generate and then adopt a block. The block can be generated or offered by a node in the network. It encodes a number of transactions (for example, in cryptocurrency, it can be monetary transactions between accounts). Furthermore, a key step is the adoption of the proposed block/related transactions by network participants, a process called consensus building. Once a block is accepted, it becomes part of the blockchain and the newly created blocks are cryptographically linked to it. After a period of time (depending on the consensus algorithm used), the block becomes a permanent part of the blockchain, i.e. it reaches finality. However, this finality does not exclude the existence of a small statistical probability that a block can be changed (intentionally by design or due to an attack), although with the addition of each new block, and in an established blockchain system, this probability becomes negligible.

Let us summarize the key points of blockchain technology in relation to the work of multi-agent robotic systems (Afanasyev & Kolotov et al., 2019):

[2] "Solving the blockchain trilemma: Decentralization, security scalability". https://www.coinbureau.com/analysis/solving-blockchain-trilemma/. Accessed: 2019-07-07

- A blockchain-based database is a database of addition only. Once the data is included in the database, they cannot be changed. The database forms the blockchain state and is distributed among the nodes.
- Each node is an agent on the blockchain network and stores a complete copy of the database. The node is responsible for transferring all incoming data received from another node to all its neighboring nodes and can generate records for changing the blockchain state. All nodes are connected through peer-to-peer communication channels. Some nodes should play the role of validation.
- Validators check the correctness of records for changing the state of the blockchain and approve them (for example, combining the records into blocks, linking blocks together, and sending new blocks to neighbors). Only verified entries are applied to all nodes to build the current blockchain state.

We wish to emphasize the advantages of blockchain technology for multi-agent systems (Afanasyev & Kolotov et al., 2019):

- Data availability is achieved through multiple duplications of data and communication.
- Consistency of data is achieved through data validation and strict rules of change compliance.
- There is no way to remove or change the data stored in the blockchain.
- Economic or reputational incentives force nodes not to violate the validation rules.

Research relevance centers on the development of distributed multi-agent robotic systems that can effectively perform different operations and independently coordinate task allocation within the system. Information exchange during the interaction of robots has particular importance in achieving the goals of a multi-agent system under conditions of uncertainty, external interference, environmental changes, or the presence of intruders have special value in the maintenance of data integrity, resilience, and security. To this end, the use of blockchain technology in multi-agent robotic systems aims to solve the problems of information exchange for a group of robots, provide a record of the interaction history, and validate task execution, enhancing the efficiency of the whole system and extending the capabilities of MARS applications.

This paper extends the study presented by the authors in Afanasyev & Kolotov et al. (2019), adding new material related to blockchain-based multi-agent systems, including aspects of implementing MARS via wireless sensor networks (WSN), ensuring integrity and security using distributed ledgers, and the prospects of deploying these systems in smart buildings, smart cities, and Industry 4.0.

The rest of this paper is structured as follows. Section 2 introduces the present state of scientific and engineering development in blockchain-based multi-agent systems. Section 3 describes and classifies typical cases identified as amenable to the application of blockchain-based robotics. Section 4 considers aspects of MARS realization using wireless sensor networks. Section 5 discusses multi-agent robotic systems related to smart buildings, smart cities, and Industry 4.0. Finally, we summarize the strong and weak sides of blockchain-based MARS and discuss the barriers that technology must overcome in order to prove its viability and become ubiquitous in Section 6.

2. Related Papers

In this section, we analyze state-of-the-art publications and applications where blockchain technology is being used for distributed multi-agent systems, with a special focusing on robotics.

A recent research series has focused on the use of blockchain technology in a shared knowledge and reputation management system to study the collective behavior of robots (Kapitonov et al., 2017; Ferrer, 2018; Strobel et al., 2018; Zikratov et al., 2016; Danilov et al., 2018; Teslya & Smirnov, 2018; Kashevnik & Teslya, 2018).

Zikratov et al. (2016) presents a trust management model for decentralized robotic networks that focuses on access control and reputation management for each node. This model provides group access based on a robot-oriented trust value that is selected and dynamically updated over time. Analysis of the system was carried out by compromising a robot through attacks.

The idea of using blockchain technology to solve security problems in multi-robot systems has been discussed in Ferrer (2018) and Danilov et al. (2018). Ferrer states that combining peer-to-peer networks with cryptographic algorithms allows a group of agents to reach an agreement (and following this, the agreement is recorded in a verifiable manner) without the need for a controlling authority. He describes some blockchain-based innovations that could provide a breakthrough in MARS applications:

- New security models and methods for preserving data confidentiality and robot entity validation;
- The design of distributed decision making and collaborative missions using special transactions in the ledger that allow robotic agents to vote and reach agreements;
- The development of blockchain ledgers to make use of different robot parameters, corresponding to changing environments without any changes in the control algorithm, thus increasing the flexibility of the robots without increasing the complexity of the MARS design;
- The creation of infrastructure for MARS to follow certain legal norms and safety rules, adopted for human society and that could even result in the building of new business models for the operation of MARS.

In Strobel (2018), the theoretical concept of managing security problems in multi-robot systems using blockchain technology is reinforced by the implementation and proof-of-concept for controlling Byzantine robots. The authors developed an approach using decentralized programs based on smart contracts to create secure swarm coordination mechanisms, as well as identifying and eliminating Byzantine swarm members through collective decision making.

Kapitonov et al. (2017) focuses on the organization of the blockchain protocol for multi-agent coordination and the control of unmanned aerial vehicles (UAVs). The paper by Danilov et al. (2018) concerns the blockchain consensus protocol, which uses an additional procedure to verify liability execution so as to prevent payment transactions to questionable service providers. For this purpose, the liability execution for agent-based service providers in the decentralized trading market is verified by a formal model checker. As proof-of-concept, an application was implemented: a taxi was modeled with a subsequent delivery check at the end of the completed mission. Lopes et al. (2019) proposes a modular architecture, combining the RobotChain (Ferrer et al., 2018) framework as a decentralized ledger for registering events with robots, smart contract technology for managing robots, and Oracle for processing all data types.

Such modular architecture can be used in various contexts (manufacturing, networks, or robot management, etc.) since it is easy to integrate, adapt, maintain, and expand into new domains. What is more, this architecture allows the refusal of tokens to accelerate the validation process or their replacement with a reputation system for managing tasks and reaching consensus, since the monetary value may not make sense for private blockchain-based networks. Examples of robotic applications include:

- Task allocation between robot networks;
- Information support of robots during operation (for example, one robot cannot recognize objects, while others can);
- Assessing robot productivity or detecting performance problems;
- Consensus voting for swarm robotics.

To ensure the interaction of heterogeneous robots in cyber-physical space, an ontology can be used that describes the knowledge and competencies of the robots in the system, provides a quick exchange of information between coalition members, and smart contracts for the allocation of sensory, computational, control, and service tasks between intelligent robots, embedded devices, and information resources (Teslya & Smirnov, 2018). To do this, Teslya and Smirno (2018) presents a methodology for creating cyber-physical smart spaces with instructions on how to create and manage coalitions of intelligent robots using knowledge processors and information stored in the blockchain. In a similar fashion, Kashevnik and Teslya (2018) discusses the cyber-physical-social system, which combines smart space technology and the blockchain. The interaction between mobile robots and humans is related in an ontology-based publication/subscription mechanism, where all data exchange is controlled and key information is stored in the blockchain network.

Intelligent cyber-physical systems can be implemented as multi-agent systems with the ability to schedule tasks by agents (Shukla et al., 2018). In such multi-agent systems, the plan execution protocol should lead to the proper completion and optimization of actions, in spite of their distributed execution. However, in unreliable scenarios there is a reasonable probability that agents will not follow the protocol due to failure or malicious execution, resulting in failure of the plan. To prevent such situations, the plan can be executed by agents through smart contracts, ensuring that the task is performed even in an untrusted environment (Shukla et al., 2018). Moreover, smart contracts can be automatically generated from manufacturing plans, resulting in the automation of the entire system with seamless integration of agents into a single cyber-physical system (Shukla et al., 2018).

Similarly, self-sustaining cyber-physical environments are formed in which all critical aspects at both the cyber level and the physical level are effectively stimulated, coordinated, and supported using blockchain-based mechanisms and protocols for data storage, communication, and coordination (Skowronski, 2019).

The advantages of a properly developed blockchain framework for organizing multi-agent systems (which can be extended to both robotic and cyber-physical systems) should meet the following requirements (Skowronski, 2019):

- Decentralization of data storage, increasing the rigidness of the system;
- Scalability and ease of joining for new agents;
- Participants can reach a consensus under conditions without trust;
- The ability to maintain trust among initially unknown agents;
- Transparency and immutability;
- Agents should remain fully autonomous and fully control their identity and private keys;
- Blockchain data has to be complete, consistent, and accessible.

3. The Classification of Blockchain-based Robotics Applications

In this section, we classify blockchain-based robotic multi-agent systems, as presented in the literature review. Let us consider and discuss our vision of possible robotic multi-agent applications based on blockchain technology, as shown in Figure 1 (this extends the classification presented in Afanasyev & Kolotov et al., 2019).

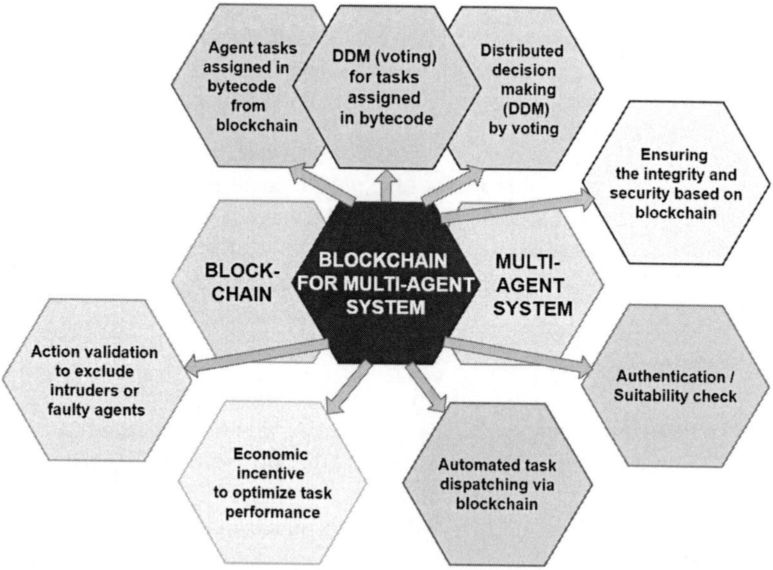

Figure 1. Classification of typical cases for the use of blockchain technology in multi-agent system robotics applications

3.1. Agent Tasks Assigned in Executable Code from the Blockchain

Let us consider a scenario in which there are multiple agents (robots), without taking into account the hardware platform. Every agent has a pre-installed control program, but they are configured to receive external executable code (bytecode), which is a set of commands in the package to implement operations and achieve the multi-agent system's goal. The word "bytecode" refers to the platform-independent executable code that results in the same sequence of command execution for any agent. As such, at the moment of sending the bytecode, there is no need to know the exact hardware platform of the agent (robot), which must execute the commands specified in the bytecode.

The blockchain can be used here to distribute bytecodes to agents (Afanasyev & Kolotov et al., 2019):

- The system that generates tasks should not be directly connected to the agent. The peer-to-peer network is used to deliver the message.
- A message will be delivered even if an agent is turned off.

- The agent is able to constantly inform about its state changes (e.g., “moved forward 10 meters”, “picked up an object”, etc.). The agent’s state is stored in the blockchain where it can be quickly recovered if the agent is turned off.
- Delayed bytecode execution can be scheduled.
- Two or more command sequences can be automatically queued up for execution by the agent.

3.2. Distributed Decision-making by Time-limited Voting

Distributed decision-making (DDM) is a challenging task in multi-agent robotic systems. A solution for using the blockchain in SWARM systems has been proposed by Ferrer (2018). This solution is based on the idea of sending cryptocoins as prizes to some addresses. However, the development of blockchain technology has seen new opportunities that effectively solve this problem; for example, the use of smart contracts in Ethereum. For this purpose, smart contract(s) should be developed to create an infrastructure by conducting polls with complex behaviors, such as time-limited voting or vote delegation.

3.3. Distributed Decision-making for Tasks Assigned in Bytecode

By combining the approaches in (3.1) and (3.2), it is possible to obtain another interesting solution, where the use of smart contracts for agents may contain tasks formulated in bytecode. Other agents may vote for actions, defining a co-generated script that can be obtained from a smart contract.

3.4. Action Validation to Exclude Intruders or Faulty Agents

Agents can be used to check each other’s actions, locations, or states. Let us look at a swarm where agents execute some actions to achieve a common goal. Periodically, agents send telemetry information on sensor measurements and locations based on odometry. Sometimes an agent may start working incorrectly and send incorrect data. Information obtained from other agents can be used to reach a consensus that the agent is wrong and the recovery procedure can begin. In this case, further co-evolution scenarios can be applied. A consensus based on information obtained from other agents can also be used to identify a robot that is behaving incorrectly; for example, if it has been hacked or infected by an attacker. To solve the performance problem of validators, the Sharding approach can also be used here (edChain, coinbureau, 2019). Data from agents are combined depending on their location forming a separate shard. Validators are coordinated for the shard and therefore the volume of information for processing is significantly reduced.

3.5. Economic Incentive to Optimize Task Performance

The financial side of the blockchain can be used as a basis for stimulating a multi-robot system. Researchers at Carnegie Mellon University performed multi-robot mapping with a market focus, coordinating a robot team and maximizing information acquisition at a minimal cost (Zlot et al., 2002). This approach demonstrated reliability and adaptability to a dynamic environment, even with the loss of colony members, in addition to its ability to withstand communication losses and disruptions. The researchers found that the market architecture in the negotiations of robots significantly improved the robot team's efficiency for environmental study (Zlot et al., 2002). Although the algorithm was designed to minimize the distance traveled during the mission, minimizing the exploration time also showed encouraging results in terms of quickly surveying terrain. This approach also allows dynamic changing priorities to minimize resource consumption for the robot team, fulfilling the multiple mission objectives for the robot group (Zlot et al., 2002).

3.6. Automated Task Dispatching via Blockchain

A blockchain-based distributed consensus can be used to dispatch, assign, and execute tasks between competing agents. In this case, the dispatching code can be written in the form of a smart contract stored in the blockchain for the implementation of the following possible scenarios:

1) The client sends a request for task execution to the smart contract dispatcher.
2) The dispatcher notifies agents about the new request.
3) The agents agree to the performance of the task in the blockchain via a peer-to-peer network.
4) Blockchain validators determine the order of agreements in accordance with the commission that a particular agent pays for processing the agreement.

5) The first agreement received by the dispatcher is confirmed by the smart contract code and the order details are provided to the appropriate agent.

As a result, the market will regulate the choice of agents that are sent to complete the task. However, certain agents can use strategies that allow them to pay more to be selected as validators, which may lead to them appearing to be the most efficient and stable service providers. A case study on the creation of such a multi-agent system is presented in the paper by Danilov et al. (2018). In this case study, an automated dispatching taxi script with validation of obligation fulfillment by comparing the route traveled to a map from a validator is implemented.

3.7. Authentication/Suitability Check

There are critical situations associated with the occurrence of risks/threats that require authentication:

- When agents do not trust each other, but use a common physical resource;
- When agents may be attacked by third parties. Hacker attacks can reveal confidential information and/or locations, change the task, and influence the outcome of the agent's activity.

In Afanasyev and Kolotov et al. (2019), the example of a service station for the replacement of electric vehicle (EV) batteries was considered. A blockchain-based solution was proposed to provide battery authentication services. Since the smart contract code in the blockchain is not changeable, the battery amortization state is available to all participants in the deal. Therefore, the service station and EVs can connect to any blockchain node to check the battery, excluding data replacement during a man-in-the-middle attack.

A similar system was described in Hua et al. (2018), where blockchain was used to store information about a battery's lifecycle and solve the problem of its replacement for a transaction without trust between the parties. The calculation of the battery price and the exchange of digital currency between an EV owner and the station, as well as the key logic, were implemented using smart contracts to solve the problem of a lack of trust. Similarly, Strugar et al. (2019) introduced an autonomous charging architecture and a billing framework for EV charging based on IOTA technology that provides resistance to hacker attacks and preserves confidential user information.

3.8. Ensuring Integrity and Security Using Blockchain

One of the key features of multi-agent robotic systems is the increased requirement for security under conditions of limited energy consumption by the devices; high productivity requirements; the physical characteristics and mobility of devices; and their compatibility with each other. The presence of both software and hardware/software components interacting with each other in multi-agent robotic systems and their environment, as well as possible variability of the cyber-physical environment, determines the susceptibility of such systems to specific types of attack.

To meet today's challenges, it is necessary to develop an integrated security approach for robotic systems (Desnitsky et al., 2016). The comprehensiveness of the approach given here relates not only to the union of various security systems; it is also very important to take into account the protection of the security system itself against attack. Blockchain can be a very effective solution here because of its reliability.

There are three basic security concepts: confidentiality, integrity, and availability. Cryptographic protocols, encrypted storage, etc. are usually used to ensure confidentiality. To support intruder detection and prevention systems, backup communication links, etc., are usually used. To ensure integrity, checksums or digital signatures are widely used. However, the use of blockchain technology can significantly improve the efficiency of systems that ensure the integrity of stored and transmitted data.

Let us consider the main areas of distributed multi-agent robotic systems where blockchain can be used (Ferrer, 2018).

- Robot sensors (Moinet et al., 2017);
- Robot storage (Shafagh et al., 2017);
- Robot architecture (Desnitsky et al., 2012);
- Interaction of agent with neighboring agents by IoT protocols (Reilly et al., 2019);
- Interaction of agent with distant hosts or the cloud through the Internet (Fotiou et al., 2018);
- Data aggregation, analysis, and cloud storage (Xu et al., 2018).

As an example, to enhance the security of interaction in multi-agent robotic systems, a new possible form of internet architecture, named data networking (NDN), can be used as a part of the system. NDN is made up of two basic things, i.e. sending a request and receiving a data packet. Regarding security issues, NDN can use a number of approaches along with blockchain to protect itself against various threats.

There are many notable types of attack (Abdallah et al., 2015), e.g., Interest Flooding, Cache Misappropriation, Data Fishing, Selfish Attack etc. It is hard to avoid these security attacks using existing security solutions due to the decentralized and dynamic characteristics of NDN. But the decentralized blockchain approach (Sharma et al., 2017) can be used to meet NDN security requirements. NDN can use the hash of interest or data by smart contract, decreasing the chance of user privacy breaches because both data identifier and user identifier are replaced with temporary names. In addition, nobody will be able to change or delete the fields due to the hashed chain blockchain structure (Kshetri, 2017).

It has already been proven (Zhu et al., 2019) that malicious actions can be detected in the NDN blockchain model and the combined blockchain and smart contract setup can play an effective role. The blockchain maintains the trajectory of interest and data in a hashed manner. Distributed blockchain for NDN contain a series of blocks; each block contains a hashed transition set. Each block has a head pointer (except the initiating block), which is linked to a previous block. This consists of a timestamp, recording the time when the block was written, and being linked to a successor block. Each block keeps a hash value composed of several hashed interest or data transition records.

4. Realization of a Multi-agent Robotic System via a Wireless Sensor Network

Wireless sensor networks (WSN) are widely used for the practical implementation of multi-agent systems, and the addition of mobile robots to the WSN structure is an acknowledged trend (Ahmed et al., 2017). Robots that are active doers in a multi-agent robotic system (MARS) can provide flexibility when installing network sensors and realizing active data acquisition, since they can perform various operations and interact with the environment (Ahmed et al., 2017; Afanasyev & Mazzara et al., 2019). Although these interactions can be predetermined or based on real-time observations, the choice of a suitable communication protocol for a robotized

WSN can be a challenge, considering the complexity and the multiple components of robots, as well as the type of communication implemented in MARS: one-to-many. One possible solution is to use the HTTP protocol over a WiFi connection, although it is not particularly suitable for bidirectional communication due to difficulties such as specifying ports and, sometimes, IP addresses for each network component; large packet sizes; high power consumption; and transmission problems for control commands via an Internet connection (Ahmed et al., 2017). The alternative solution is to use cloud computing and the Internet of things (IoT) technologies for organizing communication between nodes and controlling the robotized WSN components, especially when using the message queue telemetry transport (MQTT) protocol. Due to the small packet size and "publish/subscribe" concept, managing the connection between network devices can be made simpler and more feasible.

Currently, 6LoWPAN networks with the MQTT protocol are becoming a good solution for MARS applications due to their low power consumption, IP-driven nodes, and support for large mesh networks. Ahmed et al. (2017) proposed a robotized WSN suitable for solving various problems of environmental monitoring. This robotized WSN is an adaptive system in which intelligent agents act as moving sensors for detecting and tracking areas and the monitored environment parameters are different from certain threshold values. The iRobot Create and KUKA youBot mobile platforms with additionally-installed single-chip Gumstix Verdex pro TM XL6P computers and various expansion modules as robotic agents (mobile network nodes), were used experimentally. The functionality of the proposed robotic WSN with data exchange using 6LoWPAN is verified using the MQTTBox platform. This enables MQTT clients to be built for publishing or subscribing topics, configuring MQTT virtual device networks, and testing MQTT devices etc.

However, security is a critical issue when applying the IoT concept to organize communication between robotized WSN nodes. A successful solution to the problem of interaction between nodes, as well as recording the interaction history and performing verification, is offered by blockchain technology (see Section 3.8). This can increase the efficiency of a robotized WSN and expand the possibilities of its application.

The notion of a blockchain is associated with a publicly available chronological database of transactions recorded by a network of agents (e.g. swarm of agents). It is obvious that software robots (holons) are suitable for this purpose. Research with robotic swarms has been undertaken by Ferrer (2018). Each agent possesses private and public keys that are used to prove the origin and/or encrypt messages. As stated in Ferrer (2018), the proof may delay task execution for up to 10 minutes. In order to improve this situation, it is reasonable to use other cryptocurrencies instead of Bitcoin, for example, those based on IOTA or Ethereum. On the other hand, not all transactions should be included in the blocks. In many cases, different preprocessing schemes should be applied to deeply model the environment and classify the situation (Jotsov, 2016). Such an example is shown in Figure 2, where six security-based ontologies are depicted and their combination defines dangerous processing nodes (in red), warning zones (in yellow), and safe (uncolored) zones. Blockchain-based solutions are required only for red and sometimes yellow zones. Information preprocessing allows an increase in system efficiency and reduces its vulnerability. Other applications for modeling and data processing, also suitable for blockchain agents, are discussed in Jotsov (2016).

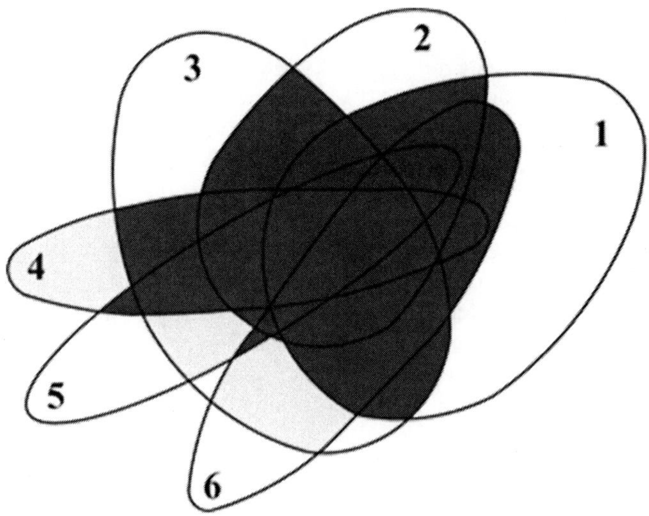

Figure 2. An ontology-related preprocessing scheme with six security-based ontologies and their combinations, identifying dangerous processing nodes (red), warning zones (yellow), and safe (uncolored) zones

5. Smart Buildings, Smart Cities, and Industry 4.0

The increasing integration of devices and robots with Internet infrastructure and everyday activities is leading to a vision of the future where the Internet itself progressively disappears from view, i.e. the conscious actions necessary to connect to the network and transfer packets of data on it are becoming more and more transparent. A future citizen will most likely not refer to the *Internet* as often as we do, similar to the way that today we simply connect devices in a stable manner very differently from the way we used to do so two decades ago, when more conscious effort was necessary and the connection could be lost several times over a work session. The more the Internet enters everyday life, the less we notice it, and the more pervasive it will become in every aspect of our personal and professional lives. For example, the use of a multi-agent system with auditable blockchain voting helps make the voting records transparent and unchangeable, enabling the important e-voting problem of how to increase the level of respondents' trust in the electronic voting system to be overcome (Pawlak et al., 2018).

This progressive integration is seen in what we call today the *Internet of things* (IoT) (Alam et al., 2017), where every object is transparently connected to the network and can communicate with other objects, systems, or individuals. In this developing scenario, robots also play a role; robotics as a discipline cannot be considered a completely separate and independently developing domain. The *Internet of robotics things* (IoRT) (Ray, 2016; Simoens et al., 2018; Batth et al., 2018; Mahieu et al., 2019; Afanasyev & Mazzara et al., 2019) is a recently defined concept, which aims to describe the integration of robotics technologies in IoT scenarios. Multi-agent robotic systems, as described in this paper, present a notable application scenario for which IoRT can constitute the working infrastructure (Gautam & Mohan, 2012), both for domestic (domotics/smart buildings) and industrial (Industry 4.0) use.

Casino et al. (2018) and Wang et al. (2019) examine the integration of blockchain technology and the Internet of things, which are expected to transform human life and provide great economic benefits. The main restrictions on such integration are insufficient data security and the issue of the level of trust.

Regarding smart buildings and software-defined buildings (SDB), blockchain is destined to represent a persistent infrastructure with pervasive application in everything concerning home resources and environmental processes (Mazzara et al., 2019), ranging from energy management to billing, and from environmental comfort to safety and surveillance. Developing the idea of smart building systems, researchers are seeking to integrate subsystems, such as intelligent networks, services, buildings and household appliances, into models of smart spaces, and even smart cities. Blockchains are being used to effectively exchange data in interacting subsystems, make connections, and operate through remote control to achieve a better life quality, greater sustainability and energy conservation, and the development of novel socio-economic systems (Lazaroiu & Roscia, 2017; Mazzara et al., 2019). The IoRT in the domotics context can act as an enabling infrastructure for blockchain-based multi-agent robotic systems. In this context, and as it applies to smart cities, which are aggregations of smart buildings, a large number of sensors collect data with high variability in accuracy, reliability, and frequency. As such, blockchain technology could permit the management of immutable public ledgers that track all activities and determine those that are more trustworthy and those that are less so, and act accordingly. This hardware and software infrastructure will simplify the interaction of different agents in different buildings, allowing for the traceability of data collection and enhancing trust, security, and accuracy in the cooperation between multiple agents.

Ecological and environmental monitoring is a sphere where all the advantages of blockchain are needed; usage of IoT and IoRT is a good way to create a big independent sensor network. Transparency, immutability, and security are all in demand for environmental monitoring. The peer-to-peer approach allows for a cheap connection between the new sensor and the global network and the public provision of environment data. Right now the concept of citizen observatories is considered one of the most suitable for blockchain technology. Projects such as WeObserve (Berti Suman & Van Geenhuizen, 2019) and WeSenseIt (Lanfranchi et al., 2014) demonstrate how citizen sensor networks help improve the comprehensiveness of ecological information; with peering technologies this can be a scalable solution.

In the context of Industry 4.0 (smart factories[3]), and with the increasing trend of automation, similar considerations on the efficacy of IoRT and blockchain hold. Blockchain technology may eventually become a pillar of a business or organization thanks to better contract management, effective

[3] Forbes, "Why Everyone Must Get Ready for the 4th Industrial Revolution". https://www.forbes.com/sites/bernardmarr/2016/04/05/why-everyone-must-get-ready-for-4th-industrial-revolution/#69263d403f90, 2016. Accessed: 2019-07-07

quality control, better accountability, and recognition and authentication of IoRT devices. In general, blockchain technology in Industry 4.0 gives us the chance to innovate and refresh the concept of cybersecurity, offering a mechanism by which activities can be immutably tracked and pseudonymized (Fernandez-Carames & Fraga-Lamas, 2019).

The progress in various technologies and their cooperation in robotics, automation, IoT, big data processing, cloud computing, and blockchain can lead to a fourth industrial revolution, with the interaction of smart factory components within a company and external industrial IoT systems providing trust and reliable control over resource distribution and products (Teslya & Ryabchikov, 2017; Kapitonov et al., 2018). For example, in logistics, where the supply chain is a multi-agent system in which each supplier has its own behavioral model and purposes, the blockchain can bring necessary transparency and trust, speeding up processes and eliminating many shortcomings of current supply chains (Casino et al., 2018; Casado-Vara et al., 2018).

6. Conclusions and Discussion

Currently, an approach consisting of the organization of an immutable distributed database storing all relevant information and providing access to agents of a multi-agent robotic system (therefore expanding the capabilities of the system as a whole) is of great relevance in the context of the proposed "Fourth Industrial Revolution". The key component of this approach is distributed ledger technology (blockchain), which allows agents to interact or allocate tasks through responsible "smart contracts". On the one hand, the reliability of the agent is primarily determined by the reputational model, allowing the agent to determine the trust level only after the fulfillment of agreed obligations. On the other hand, the automation of obligation fulfillment by an agent can provide a verification procedure that allows the verification of liability execution, even among initially unknown agents and reach a consensus in conditions without trust (Ferrer, 2018; Danilov et al., 2018; Skowronski, 2019). In addition, the goals for blockchain implementation in a multi-agent robotic system may include an increase in the interaction efficiency between agents by organizing more trusted information support; assessing robot productivity or detecting performance problems; voting consensus for swarm robotics; plan scheduling and task allocation; and deploying distributed decision-making and collaborative missions.

The basic message of this paper is on how to bring together and provide, as a guide, ideas on how frameworks, architectures, and structures supported by blockchain solutions can be used to solve practical problems faced by multi-agent robotic systems and cyber-physical systems. Relevant studies show that blockchain can play a significant role in the development of systems and applications with many agents (robots), in which the development of strategies for their coordination is conducted in such a way that agents can effectively perform their operations and intelligently coordinate task allocation (Basegio et al., 2017). An analysis of recent publications has allowed us to identify and classify groups of tasks for multi-agent robotic systems based on blockchain technology. This classification is the main contribution of this paper. Blockchain technologies can be used to expand the existing number of platforms and libraries used by researchers, or to motivate them to use a common solution that is widely distributed and tested, rather than trying to develop their own software solutions to cover similar scenarios.

Real-world scenarios may require the use of disparate agents and the performance of tasks with different structures, constraints, and complexity. Therefore, the requirements for the quality of communication in decentralized systems are increasing, including such important functions as maintaining resiliency, data integrity, and security in accessing data. Therefore, the introduction of blockchain technology for the interaction and coordination of multi-agent robotic systems is a reasonable solution for many modern research and industrial tasks.

Based on relevant investigations, the authors conclude that, at present, one of the most promising tasks in the field of developing multi-agent systems is the development of methodologies, models, structures, architectures and methods, with the aim of integrating blockchain technology with high complexity systems, such as cyber-physical systems, robot swarm, the Internet of things, the Internet of robotic things, smart buildings, smart factories, and smart cities. To fully implement such integration, it is necessary to automate and solve subtasks of different difficulty levels, such as intellectual support for agent interaction; task and plan allocation; analysis of task performance and liability execution; evaluation of agent performance; identification of improperly functioning agents and intruders; security-based issues and many others.

Many tasks remain to be done at the moment and require solutions, but a number of these are both important and urgent (Afanasyev & Kolotov et al., 2019). These include:

- The development of a conceptual model of information support for robot network during task performance;
- The design of a typical ontological model of a multi-robot network;
- The development of requirements and formal process modeling for liability execution (Mazzara, 2010; Danilov et al., 2018);
- The design of a consensus protocol for group interaction verification before launching a task based on information from a distributed ledger;
- The development of a validation methodology for task performance by the robotic system;
- The design of multi-agent robotic system architecture;
- Automatic process reconfiguration for multi-robot systems based on real scenarios;
- Analysis of cybersecurity (Dragoni et al., 2016);
- Improvement of existing frameworks allowing for multi-agent robotic networks to perform collaborative tasks in terms of scalability, decentralization, and security requirements.

References

Abdallah, E. G., Hassanein, H. S., & Zulkernine, M. (2015). A survey of security attacks in information-centric networking. IEEE Communications Surveys & Tutorials, 17(3), 1441-1454.

Afanasyev, I., Kolotov, A., Rezin, R., Danilov, K., Kashevnik, A., & Jotsov, V. (2019). Blockchain solutions for multi-agent robotic systems: Related work and open questions. In Proceedings of the 24th Conference of Open Innovations Association FRUCT (p. 76).

Afanasyev, I., Mazzara, M., Chakraborty, S., Zhuchkov, N., Maksatbek, A., Kassab, M., & Distefano, S. (2019). Towards the Internet of Robotic Things: Analysis, Architecture, Components and Challenges. arXiv preprint arXiv:1907.03817.

Ahmed, S. A., Popov, V. L., Topalov, A. V., & Shakev, N. G. (2018). Environmental monitoring using a robotized wireless sensor network. AI & SOCIETY, 33(2), 207-214.

Ahmed, S., Topalov, A., & Shakev, N. (2017). A robotized wireless sensor network based on MQTT cloud computing. In 2017 IEEE International Workshop of Electronics, Control, Measurement, Signals and their Application to Mechatronics (ECMSM) (pp. 1-6). IEEE.

Alam, N., Vats, P., & Kashyap, N. (2017). Internet of Things: A literature review. In 2017 Recent Developments in Control, Automation & Power Engineering (RDCAPE), pp. 192-197. IEEE.

Andoni, M., Robu, V., Flynn, D., Abram, S., Geach, D., Jenkins, D., & Peacock, A. (2019). Blockchain technology in the energy sector: A systematic review of challenges and opportunities. Renewable and Sustainable Energy Reviews, 100, 143-174.

Basegio, T. L., Michelin, R. A., Zorzo, A. F., & Bordini, R. H. (2017). A decentralised approach to task allocation using blockchain. In International Workshop on Engineering Multi-Agent Systems (pp. 75-91). Springer, Cham.

Batth, R. S., Nayyar, A., & Nagpal, A. (2018). Internet of Robotic Things: Driving Intelligent Robotics of Future-Concept, Architecture, Applications and Technologies. In 2018 4th International Conference on Computing Sciences (ICCS), pp. 151-160.

Berti Suman, A., & Van Geenhuizen, M. (2019). Not just noise monitoring: rethinking citizen sensing for risk-related problem-solving. Journal of Environmental Planning and Management, 1-22.

Casado-Vara, R., Prieto, J., De la Prieta, F., & Corchado, J. M. (2018). How blockchain improves the supply chain: Case study alimentary supply chain. Procedia computer science, 134, 393-398.

Casino, F., Dasaklis, T. K., & Patsakis, C. (2018). A systematic literature review of blockchain-based applications: current status, classification and open issues. Telematics and Informatics.

Danilov, K., Rezin, R., Afanasyev, I., & Kolotov, A. (2018). Towards blockchain-based robonomics: autonomous agents behavior validation. In 2018 International Conference on Intelligent Systems (IS), pp. 222-227. IEEE.

Desnitsky, V., Chechulin, A., Kotenko, I., Levshun, D., & Kolomeec, M. (2016). Application of a technique for secure embedded device design based on combining security components for creation of a perimeter protection system. In 2016 24th Euromicro International Conference on Parallel, Distributed, and Network-Based Processing (PDP), pp. 609-616. IEEE.

Desnitsky, V., Kotenko, I., & Chechulin, A. (2012). Configuration-based approach to embedded device security. In International Conference on Mathematical Methods, Models, and Architectures for Computer Network Security (pp. 270-285). Springer, Berlin, Heidelberg.

Dragoni, N., Giaretta, A., & Mazzara, M. (2016). The internet of hackable things. In International Conference in Software Engineering for Defence Applications (pp. 129-140). Springer, Cham.

Fernandez-Carames, T. M., & Fraga-Lamas, P. (2019). A Review on the Application of Blockchain to the Next Generation of Cybersecure Industry 4.0 Smart Factories. IEEE Access, 7, 45201-45218.

Ferrer, E. C. (2018). The blockchain: a new framework for robotic swarm systems. In Proceedings of the Future Technologies Conference (pp. 1037-1058). Springer, Cham.

Ferrer, E. C., Rudovic, O., Hardjono, T., & Pentland, A. (2018). Robochain: A secure data-sharing framework for human-robot interaction. arXiv preprint arXiv:1802.04480.

Fotiou, N., Siris, V. A., & Polyzos, G. C. (2018). Interacting with the Internet of Things using smart contracts and blockchain technologies. In International Conference on Security, Privacy and Anonymity in Computation, Communication and Storage (pp. 443-452). Springer, Cham.

Gautam, A., & Mohan, S. (2012). A review of research in multi-robot systems. In 2012 IEEE 7th International Conference on Industrial and Information Systems (ICIIS), pp. 1-5. IEEE.

Hua, S., Zhou, E., Pi, B., Sun, J., Nomura, Y., & Kurihara, H. (2018). Apply blockchain technology to electric vehicle battery refueling. In Proceedings of the 51st Hawaii International Conference on System Sciences.

Jotsov, V. S. (2016). Proposals for Knowledge Driven and Data Driven Applications in Security Systems. In Innovative Issues in Intelligent Systems (pp. 231-293). Springer, Cham.

Kapitonov, A., Berman, I., Lonshakov, S., & Krupenkin, A. (2018). Blockchain based protocol for economical communication in industry 4.0. In 2018 Crypto Valley Conference on Blockchain Technology (CVCBT), pp. 41-44. IEEE.

Kapitonov, A., Lonshakov, S., Krupenkin, A., & Berman, I. (2017). Blockchain-based protocol of autonomous business activity for multi-agent systems consisting of UAVs. In 2017 Workshop on Research, Education and Development of Unmanned Aerial Systems (RED-UAS), pp. 84-89. IEEE.

Kashevnik, A., & Teslya, N. (2018). Blockchain-oriented coalition formation by cps resources: Ontological approach and case study. Electronics, 7(5), 66.

Kshetri, N. (2017). Can blockchain strengthen the internet of things? IT professional, 19(4), 68-72.

Lanfranchi, V., Wrigley, S. N., Ireson, N., Wehn, U., & Ciravegna, F. (2014). Citizens' observatories for situation awareness in flooding. In ISCRAM 2014 Conference Proceedings-11th International Conference on Information Systems for Crisis Response and Management (pp. 145-154). Sheffield.

Lazaroiu, C., & Roscia, M. (2017). Smart district through iot and blockchain. In 2017 IEEE 6th International Conference on Renewable Energy Research and Applications (ICRERA), pp. 454-461. IEEE.

Lopes, V., Alexandre, L. A., & Pereira, N. (2019). Controlling robots using artificial intelligence and a consortium blockchain. arXiv preprint arXiv:1903.00660.

Mahieu, C., Ongenae, F., De Backere, F., Bonte, P., De Turck, F., & Simoens, P. (2019). Semantics-based platform for context-aware and personalized robot interaction in the internet of robotic things. Journal of Systems and Software, 149, 138-157.

Mazzara, M. (2010). Deriving specifications of dependable systems: toward a method. CoRR, vol. abs/1009.3911.

Mazzara, M., Afanasyev, I., Sarangi, S. R., Distefano, S., & Kumar, V. (2019). A reference architecture for smart and software-defined buildings. arXiv preprint arXiv:1902.09464.

Moinet, A., Darties, B., & Baril, J. L. (2017). Blockchain based trust & authentication for decentralized sensor networks. arXiv preprint arXiv:1706.01730.

Pawlak, M., Poniszewska-Marańda, A., & Kryvinska, N. (2018). Towards the intelligent agents for blockchain e-voting system. Procedia Computer Science, 141, 239-246.

Ray, P. P. (2016). Internet of robotic things: Concept, technologies, and challenges. IEEE Access, 4, 9489-9500.

Reilly, E., Maloney, M., Siegel, M., & Falco, G. (2019). A Smart City IoT Integrity-First Communication Protocol via an Ethereum Blockchain Light Client. In Proceedings of the International Workshop on Software Engineering Research and Practices for the Internet of Things (SERP4IoT 2019), Marrakech, Morocco (pp. 15-19).

Shafagh, H., Burkhalter, L., Hithnawi, A., & Duquennoy, S. (2017). Towards blockchain-based auditable storage and sharing of IoT data. In Proceedings of the 2017 on Cloud Computing Security Workshop (pp. 45-50). ACM.

Sharma, P. K., Singh, S., Jeong, Y. S., & Park, J. H. (2017). Distblocknet: A distributed blockchains-based secure SDN architecture for IoT networks. IEEE Communications Magazine, 55(9), 78-85.

Shukla, A., Mohalik, S. K., & Badrinath, R. (2018). Smart contracts for multiagent plan execution in untrusted cyber-physical systems. In 2018 IEEE 25th International Conference on High Performance Computing Workshops (HiPCW), pp. 86-94. IEEE.

Simoens, P., Dragone, M., & Saffiotti, A. (2018). The Internet of Robotic Things: A review of concept, added value and applications. International Journal of Advanced Robotic Systems, 15(1), 1729881418759424.

Skowronski, R. (2019). The open blockchain-aided multi-agent symbiotic cyber–physical systems. Future Generation Computer Systems, 94, 430-443.

Strobel, V., Castelló Ferrer, E., & Dorigo, M. (2018). Managing byzantine robots via blockchain technology in a swarm robotics collective decision-making scenario. In Proceedings of the 17th International Conference on Autonomous Agents and MultiAgent Systems (pp. 541-549).

Strugar, D., Hussain, R., Mazzara, M., Rivera, V., Afanasyev, I., & Lee, J. (2019). An architecture for distributed ledger-based m2m auditing for electric autonomous vehicles. In Workshops of the International Conference on Advanced Information Networking and Applications (pp. 116-128). Springer, Cham.

Teslya, N., & Ryabchikov, I. (2017). Blockchain-based platform architecture for industrial IoT. In 2017 21st Conference of Open Innovations Association (FRUCT), pp. 321-329. IEEE.

Teslya, N., & Smirnov, A. (2018). Blockchain-based framework for ontology-oriented robots' coalition formation in cyberphysical systems. In MATEC Web of Conferences (Vol. 161, p. 03018). EDP Sciences.

Wang, X., Zha, X., Ni, W., Liu, R. P., Guo, Y. J., Niu, X., & Zheng, K. (2019). Survey on blockchain for Internet of Things. Computer Communications.

Xu, Q., Aung, K. M. M., Zhu, Y., & Yong, K. L. (2018). A blockchain-based storage system for data analytics in the internet of things. In New Advances in the Internet of Things (pp. 119-138). Springer, Cham.

Zikratov, I., Maslennikov, O., Lebedev, I., Ometov, A., & Andreev, S. (2016). Dynamic trust management framework for robotic multi-agent systems. In Internet of Things, Smart Spaces, and Next Generation Networks and Systems (pp. 339-348). Springer.

Zlot, R., Stentz, A., Dias, M. B., & Thayer, S. (2002). Multi-robot exploration controlled by a market economy. In Proceedings 2002 IEEE International Conference on Robotics and Automation (Cat. No. 02CH37292), Vol. 3, pp. 3016-3023. IEEE.

Zhu, K., Chen, Z., Yan, W., & Zhang, L. (2018). Security attacks in named data networking of things and a blockchain solution. IEEE Internet of Things Journal.

CHAPTER 3.2

OBJECT-BASED AERIAL IMAGE ANALYSIS: A TOOLKIT FOR OPTIMIZING EDGES OF COLOR SEGMENTS

DENIS R. KASIMOV[1]

1 - Research and Educational Center "Intelligent Computer Technologies", Kalashnikov Izhevsk State Technical University, Russia. Email: kasden@mail.ru

***Abstract*.** Due to the rapid growth in the volume of aerospace monitoring data, the development of methods and systems for automatic knowledge extraction from images is becoming ever more important. Among existing approaches to the detection and classification of objects in aerospace images, object-based image analysis (OBIA), which operates with high-level information units, i.e. regions (superpixels) deriving from image segmentation, is very promising. Due to the fact that segmentation procedures operate at a relatively low level of abstraction, they often produce distorted segments that do not correspond well to the real objects present in the image. This significantly complicates the subsequent stage of semantic analysis and decision-making. To solve this problem, a method for complex optimization of color segment edges is proposed in this paper. The peculiarity of the method is that it mainly applies to geometric principles of analysis. In the course of optimization, both individual regions and aggregates of adjacent regions are considered. The experimental study, conducted with the use of ultra-high spatial resolution aerial images in urban areas, showed that through the application of the proposed method the overall quality of building detection increases and approaches the level demonstrated by machine learning techniques. Notably, the detection of objects occurs without the use of labeled datasets, preparation of which is a very time-consuming and expensive process. In general, the study introduces a new efficient toolkit in the OBIA method for extracting meaningful segments from images.

1. Introduction

The creation of methods and systems for automatic identification of objects in aerospace images is of great practical importance due to the fact that manual labeling of images is rather slow and costly and is not capable of quickly providing decision-makers with relevant information in such areas as emergency situations, environmental protection measures, and territory development planning, etc.

Among the existing methods of object identification in aerospace images, there is a class of methods that utilize low-level features of image patches, for example: the angular difference feature (Huang, Chen, & Gong 2018); the curvature scale space corner detector (Maarir, Bouikhalene, & Chajri 2017); and the morphological building index (Jimenez, Plaza, & Plaza 2017). The effectiveness of these methods has certain limits, which are fundamentally impossible to overcome without extracting and analyzing significant high-level information contained in the image.

One of the ways to improve the accuracy of object identification is to combine different sources of information. In addition to the spectral channels of the visible range (RGB), some approaches also rely on the near-infrared channel and height information obtained from LiDAR sensors. An example of a method using several types of input data is found in the work of Qin and Fang (2014). In this approach, shadow regions are first identified using a modified morphological index. Then, connected segments are determined in the extracted shadow mask and the brightness of each segment is adjusted so that the normalized difference vegetation index (NDVI) can better detect vegetation located in the shadow. Calculation of the NDVI is based on the infrared channel data. Next, the initial coarse identification of buildings is performed by the morphological cylindrical transformation of digital surface model (DSM) height data. The resulting mask of buildings is then refined by graph-cut optimization based on superpixel segmentation. Although additional sources of information are used, the method cannot process buildings located in sloped areas. The correctness of the final identification result depends largely on the quality of the LiDAR sensors. It should also be noted that data obtained from several sources and their subsequent combination require additional costs.

Another popular trend in the field of automatic object identification in aerial images is based on the use of artificial neural networks (Golovanov et al. 2018; Goswami, Gakhar, & Kaur 2014; Guo et al. 2017). A disadvantage of machine learning techniques is the need to prepare a significant amount of training data. As a rule, neural network methods display good performance indicators only when working with the class of images for which they were originally designed and trained on. In addition, neural network methods are not able to generate structural and spatial descriptions of objects and scenes, which is an important stage on the way to understanding images.

A promising approach, called object-based image analysis (OBIA) (Blaschke et al. 2014; Chen et al. 2018), in which the decision-making process is based on the analysis of higher-level units of information, i.e. image segments (superpixels), is currently being developed. The process of object identification in the image is carried out according to the following pathway: image segmentation; calculation of segment features; and classification of segments into target object categories using the rule base. The following works all provide examples of object-based methods (Khatriker & Kumar 2018; He, Zhang, & Hua 2016; Janalipour & Mohammadzadeh 2016).

The advantage of object-based methods is especially obvious in relation to high-resolution images in which spatial information begins to play a significant role. On the other hand, the OBIA approach faces a number of difficulties. For example, some researchers who have developed OBIA solutions note that it is not possible to effectively classify objects in images using shape features (Jia et al. 2013). This can be explained by the distortion of the boundaries of regions extracted from the image. The unstable nature of color area shapes is due to the imperfection of methods of automatic image segmentation. In the field of object-based image analysis, the redundancy and inaccuracy of segmentation results remains a high-priority challenge (Lhomme et al. 2009), which significantly complicates the analysis of objects and worsens the overall quality of identification. Ignoring the shape of objects is a forced measure, rather than one predetermined by its objective uselessness. It is obvious that the information on object shape is very important for decision-making; but in the existing conditions, standard shape features do not work properly and remains difficult to create sufficiently complex ones. The result is that the OBIA approach is not yet able to fully realize its significant potential.

In Antunes et al. (2018), the segmentation method (Baatz & Schäpe 2000) was used to obtain more accurate image segments, taking into account not only the color statistics of the image, but also the geometric characteristics of the segments being formed. This process is based on the technique of region growing, which involves iterative merging of neighboring image regions, with a small distance between them in the specified feature space. The merging process is controlled by an optimization strategy, which, for example, seeks to minimize the average heterogeneity (dissimilarity) of final regions. This principle of image segmentation is also used in the well-known automated system eCognition for object-based image analysis. The described method of image segmentation allows the improvement of the quality of the resulting image segments; however, the initial conditions for the decision-making stage still remain very difficult. To obtain good segmentation results from a particular type of image, it is often necessary to apply manual adjustment of the criteria by which the segmentation algorithm evaluates the similarity of adjacent regions.

In region-growing methods, the choice of seed points has a significant influence on the results of segmentation. Subudhi et al. (2015) suggest a method of point selection focused on the preservation of edges, which slightly reduces the degree of segmentation redundancy. In this method, the edge pixels are first detected using a histogram-based method. Elimination of unreliable edge pixels is then performed using a threshold operation. The remaining edge pixels are taken as starting points for subsequent region growing. The authors note that the segmentation of the image areas with blurriness or low illumination is not performed very well, since in such conditions edge detection becomes very difficult.

In general, in the direction of the OBIA approach, the stage of image segmentation is carried out, as a rule, in a semi-automatic mode by trial and error, along with human visual assessment of the results (Khatriker & Kumar 2018).

This study is carried out within the framework of a project on the development of a cognitive approach to the problems of search, analysis, and description of objects in images based on context-sensitive strategies of analysis and reasoning modeling. The approach assumes the following stages of image processing and analysis: color segmentation of image and approximation of color segment edges; identification of color area features, in which a special role belongs to the characteristics of shapes; construction of a region-adjacency graph; generation of a hypothesis about the category of the object; refinement of the object edge; and confirmation by reason modeling. The approach being developed can be considered to be object-focused and based on image segmentation.

The purpose of this paper is to develop and study a method for color area edge optimization from the automated color segmentation of an image. The less distorted and redundant the extracted color areas are, the easier and more effective further logical analysis and the decision-making process may be.

In the proposed method, optimization of the segment edges is carried out by applying a set of techniques and geometrical constraints. Among them, the following pre and post-processing techniques should be distinguished:

1) image splitting into frames;
2) merging color areas from adjacent frames;
3) splitting color regions in relatively narrow places;
4) adaptive approximation of color region edges.

This paper investigates the impact of these operations and their parameters on the quality of building detection in aerial images based on shape features.

The method helps improve the accuracy and information value of color areas, contributing to the success of subsequent high-level analysis.

2. Stages of Image Processing

Let us consider the main stages of image processing and analysis, paying particular attention to the proposed operations of color area optimization.

2.1. Image Splitting into Frames

The same image may contain different types of terrain (forest, industrial areas, residential areas, etc.), objects of incomparable size and unique texture, as well as situations where one object is obstructed by another. An effective way of improve the accuracy of color segmentation in this case is to reduce the area of analysis. The effect of this technique is particularly pronounced with respect to large-size images. Figure 1 shows some examples of color segmentation of the same part of the image, obtained during global and local processing.

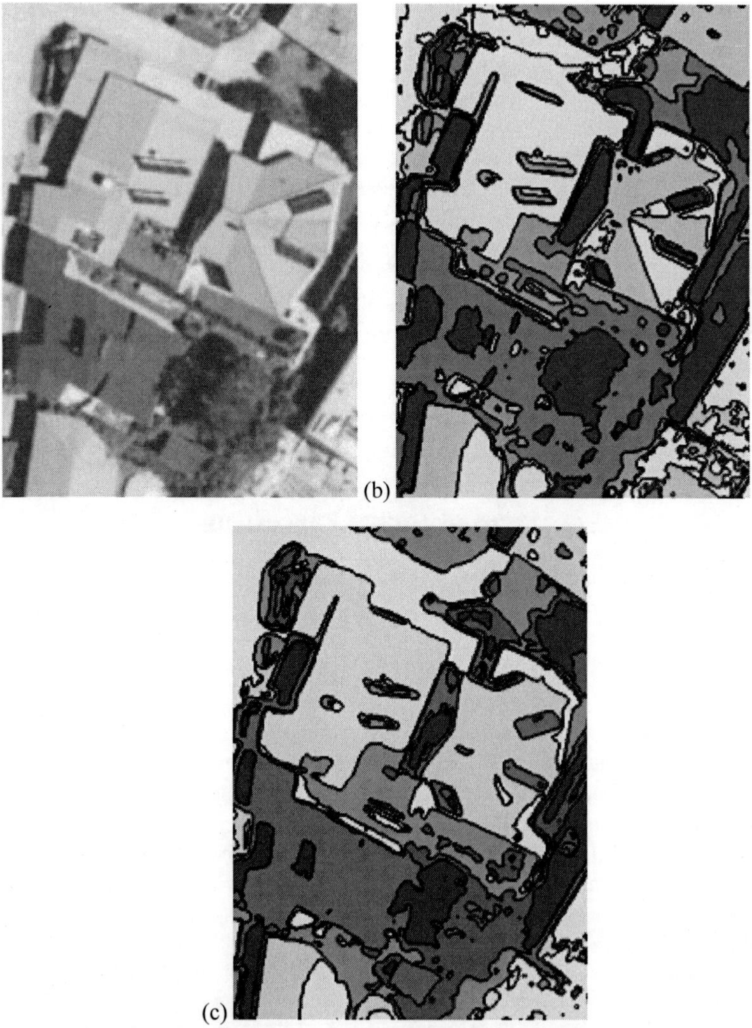

Figure 1. Comparison of segmentation results: (a) fragment of the original image; (b) segmentation in the full image; (c) local segmentation in a reduced frame

Let *W* and *H* be the width and height of the original image in pixels; *fW* and *fH* be the width and height of the bulk of the target frames; *fWm* be the minimum width of right frames; and *fHm* be the minimum height of bottom frames. Then the initial large-size image is split into a host of frames:

$$IMAGE = \{\ p_{ij} \mid p_{ij} \text{ is a pixel}, i = 1...H, j = 1...W\ \},$$

$$FRAMES = \{\ F_{mn} \subseteq IMAGE \mid m = 1...M, n = 1...N\ \},$$

$$M = \begin{cases} 1, \text{if } H \le fH; \\ H \text{ div } fH, \text{if } H \text{ mod } fH < fHm; \\ H \text{ div } fH + 1, \text{else}; \end{cases}$$

$$N = \begin{cases} 1, \text{if } W \le fW; \\ W \text{ div } fW, if\ W \text{ mod } fW < fWm; \\ W \text{ div } fW + 1, \text{else}. \end{cases}$$

2.2. Color Segmentation and Edge Detection

Each image frame undergoes color segmentation. The applied segmentation method relies on constructing a three-dimensional histogram in the color space HSV and searching for the local maximums on the histogram by scanning the color space with a three-dimensional neighborhood analysis operator (Kasimov et al. 2018).

Next, the edges of the resulting color regions are determined. A pixel belongs to the edge of a color region if at least one of the adjacent pixels belongs to a different cluster of the color palette (Kasimov et al. 2018). For convenience of further processing, the edges are represented in a vector form by cyclic lists of straight line segments, and their weak approximation is performed.

As a result of these operations, we obtain a set of color regions, each of which has a characteristic average color and an external geometric shape as a list of edge segments:

$$EdgeDetection(Segmentation(F_{mn})) = REGIONS = \{\ (A, c, E) \mid A \subset F_{mn}, c \text{ is a color value}, E = (l_1, \dots, l_k)\ \}.$$

The resulting color areas correspond in varying degrees to the real objects present in the image, or their individual parts. Edge sections are ordered counterclockwise.

2.3. Merging Areas from Adjacent Frames

The downside of splitting image into frames is the potential loss of information about objects located directly on the borders of adjacent frames. The risk of missing or misclassifying targets at the decision-making stage increases.

At the same time, a close look shows that color areas belonging to the same object and lying on opposite sides of the border of neighboring frames have adjacent sections and similar colors. Such areas can be successfully identified and merged.

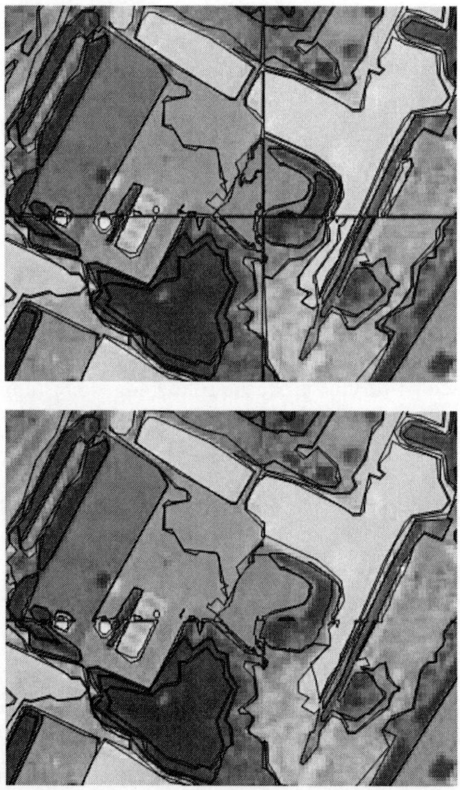

(a) (b)

Figure 2. Example of region merging: (a) source areas; (b) result areas

Figure 2 shows an example of the merging of areas that fall into adjacent frames.

Areas are merged by Algorithm 1.

Algorithm 1 Merging image regions

1) Identify the areas $AdjRegions \subseteq REGIONS$ that have vertical and horizontal edge segments at the frame edges. The proximity of a segment to a frame edge is determined while taking into account the tolerance threshold $\delta = 2$ px, compensating for deviations that may occur after the weak approximation of the region edges.

2) For each $R_1 = (A_1, c_1, E_1) \in AdjRegions$, $R_2 = (A_2, c_2, E_2) \in AdjRegions \setminus \{R_1\}$ lying in adjacent frames and having a similar color ($c_1 \cong c_2$) perform the following:

2.1) In lists E_1 and E_2, identify the alternating sequence $K = (n_1, s_2, n_3, s_4, \ldots, n_k)$, $k \geq 1$ (see Figure 3) of non-adjacent and adjacent sections located along the corresponding frame edge, based on the operation of projecting the end-points of the line segments onto each other. If one area flows exactly into another, the sections n_1 and n_k will be fictitious (of zero length).

2.2) If there are no adjacent sections, do not merge the regions.

2.3) Loop by adjacent sections (s_i):

2.3.1) Sample the pixel bands P_1 and P_2 located near the section s_i: $P_1 \subseteq A_1$, $Near(P_1, s_i)$, $P_2 \subseteq A_2$, $Near(P_2, s_i)$.

2.3.2) If the colors of the pixels at the edges of the frames are not similar: $AvgColor(P_1) \neq AvgColor(P_2)$, then do not merge the regions.

2.4) Perform the merge $R' = R_1 \cup R_2$, excluding the edge fragments that assigned to the sections of the sequence K under indices from 2 to $k-1$.

2.5) Loop over all n_i such that $1 < i < k$:

2.5.1) Create a new area R'' by merging fragments of edges E_1 and E_2, corresponding to the section n_i. Mark that the region R'' is nested in the region R'.

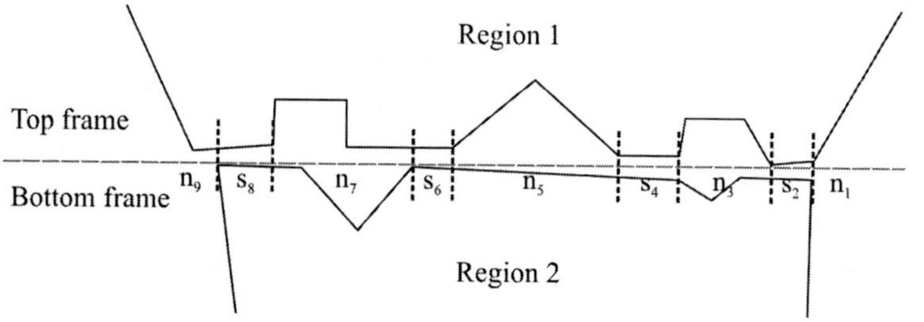

Figure 3. Example of a sequence of non-adjacent and adjacent segments

The execution of the above algorithm is repeated as long as changes are found.

To check the similarity of color values, complex color difference models can be used, as described in Sharma, Wu, and Dalal (2005) and Lv and Fang (2018). However, in the considered range of problems the introduction of several simple and natural rules of similarity is quite successful. For example, colors are considered similar if at least one of the following statements regarding their HSV components is true:

1) both colors are dark ($V_1 \leq 27 \wedge V_2 \leq 27$);
2) both colors are gray and close in brightness ($S_1 \leq 17 \wedge S_2 \leq 17 \wedge \Delta V \leq 20$);
3) the difference between the component values is small ($\Delta H \leq 30 \wedge \Delta S \leq 15 \wedge \Delta V \leq 15$).

The proposed merging algorithm may be useful not only for combining regions from different frames of the same image, but also for solving the problem of merging images of adjacent territories.

2.4. Splitting Areas in Narrow Places

Due to the fact that the task of automatic color segmentation, by definition, involves pixels grouping by the criterion of color similarity, we are faced with the well-known problem of a semantic gap (Eakins & Graham 1999). One of its manifestations is the appearance of color areas that capture objects of different semantic categories due to the fact that these objects are close to each other and their pixels have similar color values. However, it can be seen that there are many regions that connect different objects through narrow "isthmuses". As a rule, real objects do not have such "isthmuses". Thus, another method of color area refining on the basis of geometric splitting of the figure into parts in narrow places suggests itself.

According to Algorithm 2, the region *Edge* is cut into the sub-regions *NewEdges* in optimal places.

In Algorithm 2, the function *Project*(p, e) returns a point resulting from the projection of point p onto the line segment e, if any. Function *Intersect*(v, e) returns the true value if line segments v and e intersect each other. Function *SubChain*(E, p_1, p_2) returns sub-chain S of chain E, which corresponds to a counterclockwise traversal from p_1 to p_2; and S is closed by a line segment between p_2 and p_1.

The algorithm searches for a point of the chain-cut in which the ratio of the section length to the perimeter of the resulting sub-chain is minimal. The area is split only if it is not very small and there is a satisfactory (rather narrow in relative terms) cutting place.

In order to optimize the algorithm, only those cutting places are considered in which there is a negative inflection of the area edge.

Note that Algorithm 2 analyzes each region separately, looking for places where the opposite elements of the edge are close to each other. This type of splitting is intra-object. Unfortunately, this does not solve all the problems of splicing. After color segmentation of an image, it is not uncommon to come across situations where one color area lies within another, coming close to its edge. Between such areas there are narrow places, which are bridges connecting different objects. In such places it is also necessary to make a split, which in this case has an inter-object character. This type of split is performed by Algorithm 3.

Algorithm 2 Intra-object splitting

```
NewEdges ← ∅
Queue ← { Edge }
foreach E ∈ Queue do
    m ← +∞
    if Area(E) is not VerySmall then
        for i = 1 .. |E| do
            pi ← StartPoint(ei)
            for j = 1 .. |E|, j ≠ i do
                foreach pj ∈ { StartPoint(ej), Project(pi, ej) }
                    v ← LineSegment(pi, pj)
                    if ∃k (k ≠ i ∧ k ≠ j ∧ Intersect(v, ek)) then
                        continue
                    end
                    r1 ← |v| / Perimeter(SubChain(E, pi, pj))
                    r2 ← |v| / Perimeter(SubChain(E, pj, pi))
                    if (r1 and r2 are not Large) ∧ (r1 + r2 < m) then
                        m ← r1 + r2
                        p1 ← pi
                        p2 ← pj
                    end
                end
            end
        end
    end
    if m = +∞ then
        NewEdges ← NewEdges ∪ { E }
    else
        S1 ← SubChain(E, p1, p2)
        S2 ← SubChain(E, p2, p1)
        Queue ← Queue ∪ { S1, S2 }
    end
end
```

In Algorithm 3, the function *ClosestPoint*(p_1, E^O) returns a point that lies on edge E^O and is closest to the point p_1.

In this algorithm the similarity of the resulting sub-regions with respect to the perimeter value serves as the criterion of optimal splitting.

Algorithm 3 Inter-object splitting

<u>INPUT</u>: E^O – region that could be potentially divided (outer edge); E^I – region lying inside E^O (inner edge).
<u>OUTPUT</u>: $S^R{}_1$ and $S^R{}_2$ – resulting subregions.

$m \leftarrow 0$
for each $e_1 \in E^I$ **do**
    $p_1 \leftarrow StartPoint(e_1)$
    $p_2 \leftarrow ClosestPoint(p_1, E^O)$
    if $|p_1p_2|$ is *Relatively Large* **then** continue
    for each $e_2 \in E^I \setminus \{e_1\}$ **do**
        $p_3 \leftarrow StartPoint(e_2)$
        $p_4 \leftarrow ClosestPoint(p_3, E^O)$
        if $|p_3p_4|$ is *Relatively Large* **then** continue
        $S_1 \leftarrow SubChain(E^O, p_2, p_4)$
        $S_2 \leftarrow SubChain(E^O, p_4, p_2)$
        $L_1 \leftarrow Perimeter(S_1)$
        $L_2 \leftarrow Perimeter(S_2)$
        $r_1 \leftarrow |p_1p_2| / L_1$
        $r_2 \leftarrow |p_3p_4| / L_1$
        $r_3 \leftarrow |p_1p_2| / L_2$
        $r_4 \leftarrow |p_3p_4| / L_2$
        $w \leftarrow \mathrm{MIN}(L_1, L_2) / \mathrm{MAX}(L_1, L_2)$
        if (r_1, r_2, r_3 and r_4 are **not** *Large*) $\wedge$ $(w > m)$ **then**
            $m \leftarrow w$
            $S^R{}_1 \leftarrow S_1$
            $S^R{}_2 \leftarrow S_2$
        end
    end
end

The loops of splitting selection variants go through the elements of the inner chain (E^I) rather than the outer one. This is because, first of all, it consists of a smaller number of elements and, secondly, in the contrary case the algorithm would consider many unpromising variants of cross-sections that are too wide.

Figure 4 shows an example of color area splitting.

The relative thresholds used in the algorithms are set to be sufficiently large to obtain the "intense" splitting effect shown in Figure 5.

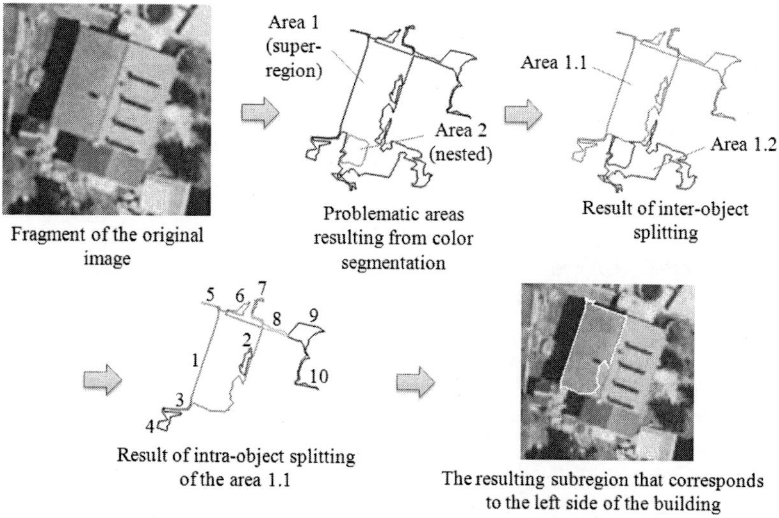

Figure 4. Refining the object edge by splitting the super-region into several parts

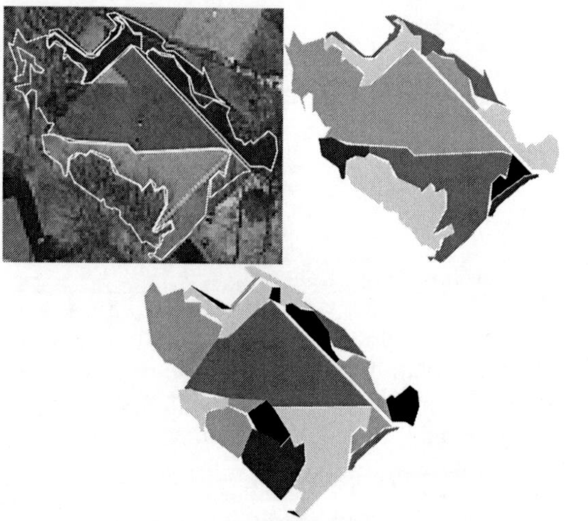

(a) (b) (c)

Figure 5. Example of intensive splitting: (a) initial color segments; (b) areas after splitting in narrow places; (c) areas after intense partitioning

As a result of intense splitting, it is possible to separate different objects from each other, even if the length of their common border is relatively large. Of course, this operation has a downside associated with an increase in the degree of object fragmentation. However, this problem is less serious than the presence of super-regions. This can be successfully solved by applying the procedure of single object assemblage from several fragments. This procedure corresponds to the search for optimal paths on the region adjacency graph and includes the following actions:

1) construction of the graph showing the adjacency of sections of region edges;
2) selection of candidate areas according to a given set of criteria (for example, small in size, rectangular, triangular, and trapezoidal);
3) determining the group (family) of adjacent areas by identifying connected components in the adjacency graph;
4) selection of subgroups that meet specified criteria (e.g., the similarity of the shades, the squareness of the external hull);
5) synthesis of a single region by constructing the non-convex hull;
6) variant assessment.

Figure 6 demonstrates the operation of the described procedure through a specific example. With this technique, the cut-off parts can be successfully returned back to the object.

Areas belonging to the category of "Greenery" rarely capture other categories of objects and therefore do not require splitting. It is advisable to exclude them from the input of the splitting algorithm. Such filtering is relatively easy to implement and saves computational resources. To identify natural background regions, it is necessary to analyze their color, texture, and shape. A region can be classified as "Greenery" if it has a green tint (in terms of the H-component of the average color); no low lightness (according to the V-component); heterogeneous texture (by the corresponding descriptor); a large area; high tortuosity (in terms of the number of changes in the direction of the edge bend); and low straightness (regarding the proportion of significant straight line segments). Of course, these criteria do not guarantee 100 % recall and precision in identifying areas of "Greenery". But they do allow the exclusion of a significant part of these areas from the resource-intensive process of splitting and subsequent assemblage.

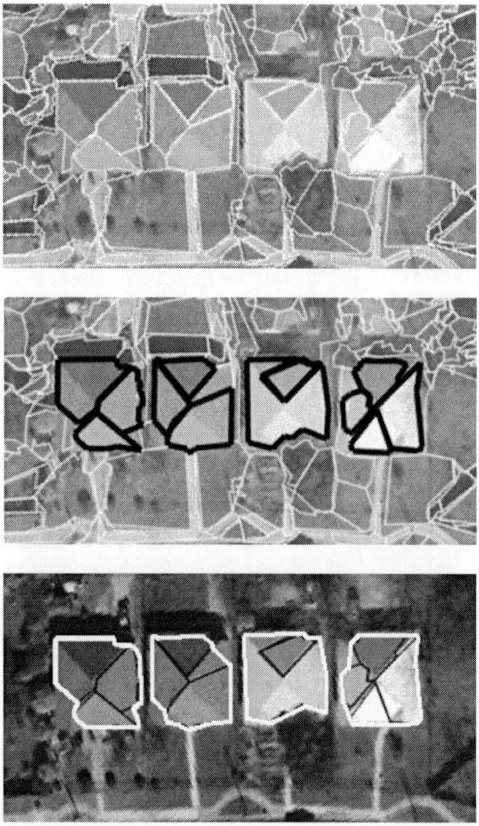

(a) (b) (c)

Figure 6. Refining the object edge by assembling from several fragments: (a) set of areas after segmentation and splitting; (b) selection of candidates according to the criteria of size and shape (rectangular, triangular, trapezoidal), determining families of adjacent areas; (c) search and evaluation of variants of subgroups in each family, construction of a non-convex hull

It is also possible to successfully identify and exclude from the splitting process the part of the regions that forms the "Road Network". To do this, the following strategy is used. First, determine the bearing (most reliable) parts of roads according to the following criteria: asphalt color; large elongation or no great average width; high straightness; and considerable size. Then iteratively consider regions adjacent to the previously identified road parts, checking the similarity between them in color and taking into account geometric constraints, such as the width of the road along its entire length. Figure 7 shows an example of this application. To capture even more fragments of the road network, it is necessary to introduce some additional steps of analysis, for example, identification of individual roads in the set of regions obtained, control of the road intersection angles, and filling of road gaps, etc.

Figure 7. The result of automatic detection of road network sections where it is not advisable to split them into parts

2.5. Adaptive Approximation of Edges

Depending on the hypothesis made about the category of the object that is being tested, different degrees of approximation of its shape-features may be required. For example, the characteristic of edge straightness is among the most distinctive features of a road. The straightness of the road edge may be reduced due to the influence of other objects (for example, due to the presence of a parked car on the roadside or an overhanging tree) and due to the fact that during automatic color segmentation of the image, the road area can be erroneously replenished with fragments of adjacent territory (sidewalks, roadside verge) that are similar in color, or, on the contrary, contaminated; or defective fragments of different color may be erroneously subtracted from it. As a result, winding fragments appear in the shape of the edge, reducing the straightness index and making it difficult to identify the road. The task becomes even more complicated by the fact that straight sections need to be identified over a long distance, because roads have a significant length in comparison to other categories of objects.

This problem can be solved by making a stronger approximation of potential road edges, as shown in Figure 8.

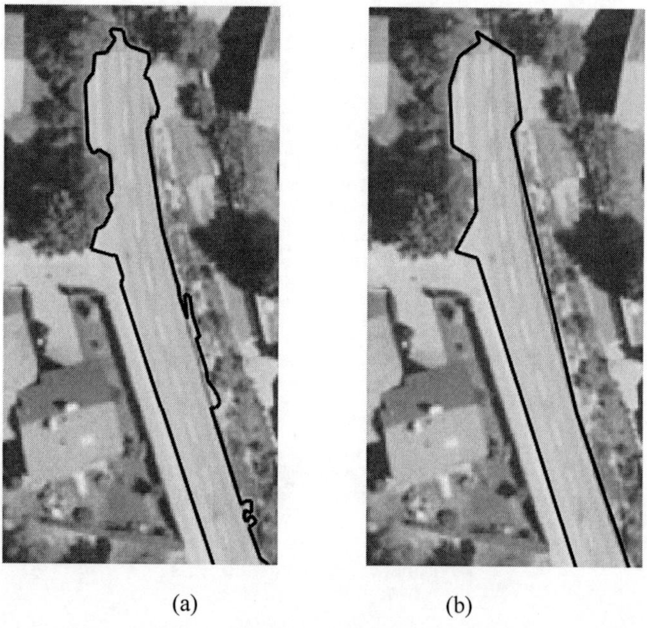

(a) (b)

Figure 8. Example of edge approximation: (a) initial boundary obtained in the course of weak approximation at the initial (general) stage of processing; (b) re-approximated boundary

The applied approximation method, considered in detail in Kasimov et al. (2018), results in the formation of straight line segments and circular arcs by recursive augmentation of approximated chains.

The threshold of maximal deviation of the resulting chain from the original one depends on the size of the area to adapt to objects of different sizes.

2.6. Feature Calculation and Object Classification

In the process of object detection, various geometric and textural features and relations between color regions are analyzed; hypotheses about categories of objects are put forward and verified by logical inference. In contrast to existing methods (Jia et al. 2013), the approach being developed here assigns a special role to the shape-features and external surroundings of the areas as elements on the basis of which reason modeling may be carried out, thereby overcoming the semantic gap and obtaining more reasonable results in identification.

At the current stage of project development, the following features are automatically calculated and used in decision-making: significant elements of the edge; straightness; the presence of three sides of a rectangle; the presence of the corners of a rectangle; area; average width; elongation; squareness; circleness; tortuosity; average color; and contour point density.

Below is an example of a classification rule that deals with the partial hiding of a building by trees:

$$\exists a,b \in REGIONS\ (Trees(a) \wedge Adjacent(a, b) \wedge HasRectAngles(b) \wedge Tortuosity(CommonBorder(a, b)) \text{ is } High) \rightarrow Building(b).$$

During logical analysis of the image, the inference tree is built.

A detailed consideration of the stage at which the classification of objects takes place is beyond the scope of this paper. The development of strategies and methods for analysis and decision-making applied at the classification stage is the subject of a separate study.

3. Experimental Study

An experimental study of the presented techniques was made on images from the publicly available IAIL benchmark dataset (Maggiori et al. 2017; Huang et al. 2018). The images have dimensions of 5,000 × 5,000 and high spatial resolution (0.3 m).

The experiments seek to assess the degree to which the proposed techniques of color area optimization can affect the quality of the object identification results. In the experiments, the target objects were buildings.

The indicators *Recall*, *Precision* and *F1*-measure, widely used in the problems of semantic image labeling (Fernandez-Moral et al. 2017), were applied as evaluation metrics. These indicators are calculated as follows:

$$Recall = \frac{TP}{TP + FN},$$

$$Precision = \frac{TP}{TP + FP},$$

$$F1 = \frac{2TP}{2TP + FN + FP},$$

where *TP* is the number of target objects successfully identified by the system; *FN* is the number of target objects missed by the system; and *FP* is the number of false objects present at system output.

F1-measure is an integrated representation of the *Recall* and *Precision* metrics. The calculation of the indicators was based on the ground truth data available in the benchmark.

Table 1 presents the obtained values of performance indicators for different sizes of image cutting into frames.

Table 1. Dependence of identification quality on image splitting size

No	Frame size	Recall	Precision	F1-measure
1	1000×1000	0.611	0.723	0.663
2	750×750	0.620	0.738	0.674
3	500×500	0.651	0.724	0.686
4	400×400	0.666	0.747	0.704
5	300×300	0.700	0.803	**0.748**
6	250×250	0.671	0.808	0.733
7	200×200	0.723	0.774	**0.747**
8	150×150	0.654	0.774	0.709
9	125×125	0.671	0.825	0.740

Thus, the optimal frame size is within the range of 200 to 300 pixels. With an excessive reduction in the frame size, the subsequent merging of color areas becomes difficult and the number of errors increases. In addition, the image processing speed is noticeably reduced.

When the technique of merging color areas from neighboring frames is not used, the value of *F1* decreases by 4 % with a frame size of 300 × 300 and by 6.7 % with a frame size of 200 × 200. The degradation in quality of the final result also takes place when the operation of splitting areas into parts in narrow places is disabled: overall deterioration by 11.2 % and 9.1 %, respectively, due to a noticeable decrease in recall.

Table 2 shows the effect of changing in the degree of region edge approximation. Best results are obtained with a sufficiently large threshold of relative deviation (2.5 % of the region size).

Table 2. Impact of approximation on the result

No	Relative deviation threshold	Recall	Precision	F1-measure
1	0 (without approximation)	0.363	0.799	0.499
2	0.005	0.380	0.911	0.536
3	0.010	0.426	0.887	0.575
4	0.015	0.509	0.864	0.640
5	0.020	0.617	0.818	0.704
6	0.025	**0.700**	**0.803**	**0.748**
7	0.030	0.734	0.732	0.733
8	0.035	0.777	0.690	0.731
9	0.040	0.800	0.648	0.716
10	0.045	0.823	0.630	0.713
11	0.050	0.834	0.606	0.702

Thus, the approximation has a significant impact on the final result of image object classification.

Since the comparison of color segmentation quality, as with other methods, presents a particular difficulty, the author provides comparative indicators of the overall performance of building identification. Detection was carried out on the basis of shape features and logical rules, as mentioned in Clause 2.6.

Table 3 contains numerical estimates of the quality of the results obtained through the use of alternative methods and the proposed method for processing images of areas of the West Tyrol from the IAIL benchmark dataset. In our case, image processing was performed with optimal parameter values determined in the above experiments. The alternative methods are based on neural networks; their performance evaluations are adopted from Maggiori et al. (2017).

Table 3. Comparative numerical evaluations of overall building identification performance

Method	IoU	Accuracy
FCN (Maggiori et al. 2017)	46.86	95.83
Skip (Maggiori et al. 2017)	54.91	96.47
MLP (Maggiori et al. 2017)	57.95	96.66
Proposed OBIA-method	53.13	85.10

The pixel-oriented metrics *IoU* and *Accuracy* that are used here as evaluation metrics are also popular in the field of semantic classification. The *IoU* index considers the sets of target object pixels in the prediction and in the ground truth and is calculated as the ratio of the cardinality of the intersection of these sets to the cardinality of their union. The *Accuracy* indicator shows the proportion of true pixels in the prediction (Fernandez-Moral et al. 2017). These indicators are characterized by the fact that they register the slightest deviation in the results produced by the program relative to the ground truth, which, of course, is of no small importance for use in cartographic tasks. This feature explains the fact that the *IoU* and *Accuracy* indicators obtained by the author are lower when compared to the above estimates of *Recall* and *Precision*, allowing a certain degree of deviation of the detected object edges from the ground-truth.

The comparative assessments show that the quality of identification is at the same level as that shown by machine learning methods based on neural networks. It should be noted that our results were obtained without using any training data. It is known that the preparation of training datasets involves significant costs. The proposed approach allows the automation of this process and simplifies it by at least half.

Figure 9 provides an example for visual comparison between detection results and reference data.

As noted, our approach mainly focuses on the shape features of objects. The experiments confirmed the viability of this strategy, but also revealed some difficulties that need to be overcome to achieve better performance:

- A rectilinear rectangular shape does not always correspond solely to buildings. Some other man-made objects (sports grounds, yard areas, parking lots, lawn areas, etc.) may also present such a shape. To eliminate misclassification of this kind, it is necessary to introduce additional spectral and texture features, as well as to strengthen the analysis of the scene using case-based reasoning.

(a)

(b)

(c)

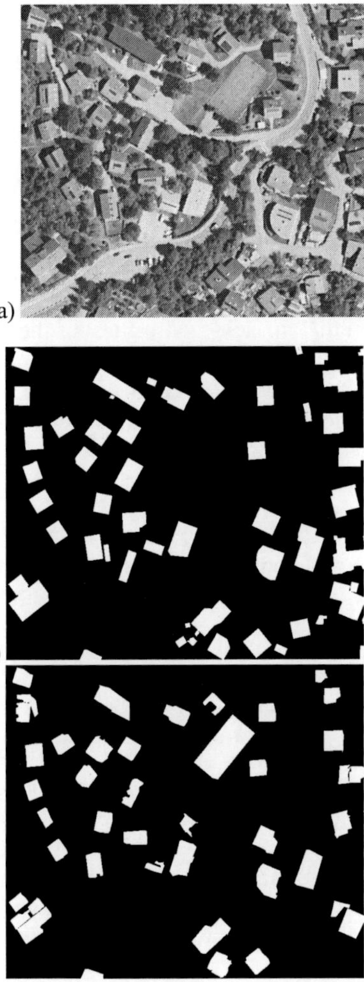

Figure 9. Results in comparison with ground-truth labeling: (a) input image; (b) reference data; (c) results of the program

- The shape of artificial objects may be distorted if they are partially obstructed by other objects (for example, trees or the shadow of large buildings). In this case, it may be reasonable to identify common parts of the edges of adjacent objects and exclude them from consideration during calculation of the shape-features so that the winding sections do not mix with the straight fragments of the object edge. This will eliminate the averaging effect.

Naturally, the spatial resolution of images and the presence of noise affect the accuracy of shape analysis. If, at lower resolutions, the clear corners of artificial objects are preserved, then in such images the proposed approach remains quite workable. When the corners are rounded, the object of interest becomes difficult to distinguish from other objects. Noise is usually removed at the first stages of image processing and is an issue. In the proposed approach, the noise is eliminated at the upper levels, with the help of logical analysis.

It should be noted that the achieved values for the quality indicators are preliminary. In particular, no attempts have been made to refine the detected objects, for which special techniques and mechanisms of changing the degree of abstraction are needed. It is expected that, after creation of adaptive context-sensitive strategies of reason modeling on spatially distributed image data, it will be possible to significantly improve the overall effectiveness of the system.

In general, the experiments have shown that the proposed techniques can improve the results of color segmentation. The correctness of the latter has a direct impact on the final quality of the identification of image objects.

4. Conclusion

In the object-based image analysis approach, the degree of correspondence between the image segments formed by the procedure of automatic image segmentation and the real objects present in the image largely determines the accuracy of the semantic interpretation of the image. The segmentation procedure operates at a relatively low level of abstraction and therefore often produces an excessive or, conversely, insufficient set of segments.

To solve this problem, a method is proposed here for the complex optimization of color segment edges. The peculiarity of the method is that it relies mainly on geometric principles of analysis. In the course of optimization, both individual regions and aggregates of adjacent regions are considered.

Experimental study has shown that with the application of the proposed method, the overall quality of object detection approaches the level demonstrated by methods based on artificial neural networks. Reaching this level of performance is a very significant result due to the fact that, unlike competing machine learning techniques, it was achieved without the use of labeled datasets, the preparation of which is both laborious and expensive.

As such, the proposed toolkit supplements the OBIA approach in terms of extracting and refining information, on the basis of which the semantic analysis subsystem makes classification decisions.

Furthermore, it aims to strengthen the algorithms of object shape analysis; increase the depth of reasoning and the degree of context consideration; develop adaptation mechanisms that will select analysis methods and inference rules depending on the type of the image, the imaging conditions, and the content of the current image area. Further research in this area will focus on these questions.

Acknowledgment

This work is supported by the Russian Science Foundation under grant No. 18-71-00109.

References

Antunes, Rodrigo Rodrigues, Edilson de Sousa Bias, Gilson Alexandre Ostwald Pedro da Costa, and Ricardo Seixas Brites. 2018. "Object-Based Analysis for Urban Land Cover Mapping using the InterIMAGE and the SIPINA Free Software Packages." *Bulletin of Geodetic Sciences* 24, is. 1: 1–17. http://dx.doi.org/10.1590/s1982-21702018000100001

Baatz, Martin, and Arno Schäpe. 2000. "Multiresolution segmentation: an optimization approach for high quality multi-scale image segmentation." In *XII Angewandte Geographische Informationsverarbeitung, AGIT Symposium. Proceedings*: 12–23.

Blaschke, Thomas, Geoffrey J. Hay, Maggi Kelly, Stefan Lang, Peter Hofmann, Elisabeth Addink, Raul Queiroz Feitosa et al. 2014. "Geographic Object-Based Image Analysis: Towards a new paradigm." *ISPRS Journal of Photogrammetry and Remote Sensing* 87: 180–191. https://doi.org/10.1016/j.isprsjprs.2013.09.014

Chen, Gang, Qihao Weng, Geoffrey J. Hay, and Yinan He. 2018. “Geographic Object-based Image Analysis (GEOBIA): Emerging trends and future opportunities.” *GIScience & Remote Sensing* 55, is. 2: 159–182. https://doi.org/10.1080/15481603.2018.1426092

Eakins, John, and Margaret Graham. 1999. *Content-based image retrieval: Technical report, JISC Technology Applications Programme*. Newcastle: University of Northumbria at Newcastle.

Fernandez-Moral, Eduardo, Renato Martins, Denis Wolf, and Patrick Rives. 2017. “A new metric for evaluating semantic segmentation: leveraging global and contour accuracy.” *Workshop on Planning, Perception and Navigation for Intelligent Vehicles, PPNIV17*. https://doi.org/10.1109/ivs.2018.8500497

Golovanov, Sergey, Rauf Kurbanov, Aleksey Artamonov, Alex Davydow, and Sergey Nikolenko. 2018. “Building Detection from Satellite Imagery Using a Composite Loss Function.” *The IEEE Conference on Computer Vision and Pattern Recognition (CVPR) Workshops*: 229–232. https://doi.org/10.1109/CVPRW.2018.00040

Goswami, Anil Kumar, Shalini Gakhar, and Harneet Kaur. 2014. “Automatic Object Recognition from Satellite Images using Artificial Neural Network.” *International Journal of Computer Applications* 95, no. 10: 33–39. https://doi.org/10.5120/16633-6502

Guo, Zhiling, Qi Chen, Guangming Wu, Yongwei Xu, Ryosuke Shibasaki, and Xiaowei Shao. 2017. “Village Building Identification Based on Ensemble Convolutional Neural Networks.” *Sensors (Basel)* 17, is. 11: 2487. https://doi.org/10.3390/s17112487

He, Yuanrong, Xinxin Zhang, and Lizhong Hua. 2016. “Object-Based Distinction between Building Shadow and Water in High-Resolution Imagery Using Fuzzy-Rule Classification and Artificial Bee Colony Optimization.” *Journal of Sensors* 2016: 2385039. http://dx.doi.org/10.1155/2016/2385039

Huang, Bohao, Kangkang Lu, Nicolas Audebert, Andrew Khalel, Yuliya Tarabalka, Jordan M. Malof, Alexandre Boulch et al. 2018. “Large-scale semantic classification: outcome of the first year of Inria Aerial image labeling benchmark.” *IGARSS2018—IEEE International Geoscience and Remote Sensing Symposium*: 1–4. http://dx.doi.org/10.1109/IGARSS.2018.8518525

Huang, Xin, Huijun Chen, and Jianya Gong. 2018. “Angular difference feature extraction for urban scene classification using ZY-3 multi-angle high-resolution satellite imagery.” *ISPRS Journal of Photogrammetry and Remote Sensing* 135: 127–141. https://doi.org/10.1016/j.isprsjprs.2017.11.017

Janalipour, Milad, and Ali Mohammadzadeh. 2016. “Building Damage Detection Using Object-Based Image Analysis and ANFIS From High-Resolution Image (Case Study: BAM Earthquake, Iran).” *IEEE Journal of Selected Topics in Applied Earth Observations and Remote Sensing* 9, is.5: 1937–1945. https://doi.org/10.1109/JSTARS.2015.2458582

Jia, Jianhua, Ning Yang, Chao Zhang, Anzhi Yue, Jianyu Yang, and Dehai Zhu. 2013. "Object-oriented feature selection of high spatial resolution images using an improved Relief algorithm." *Mathematical and Computer Modelling* 58, is. 3–4: 619–626.
https://doi.org/10.1016/j.mcm.2011.10.045

Jimenez, Luis Ignacio, Javier Plaza, and Antonio Plaza. 2017. "Efficient Implementation of Morphological Index for Building/Shadow Extraction from Remotely Sensed Images." *The Journal of Supercomputing* 73, is. 1: 482–494.
https://doi.org/10.1007/s11227-016-1890-9

Kasimov, Denis R., Aleksandr V. Kuchuganov, Valeriy N. Kuchuganov, and Pavel P. Oskolkov. 2018. "Approximation of Color Images Based on the Clusterization of the Color Palette and Smoothing Boundaries by Splines and Arcs." *Programming and Computer Software* 44, no. 5: 295–302.
https://doi.org/10.1134/S0361768818050043

Khatriker, Shweta, and Minakshi Kumar. 2018. "Building Footprint Extraction from High Resolution Satellite Imagery Using Segmentation." *The International Archives of the Photogrammetry, Remote Sensing and Spatial Information Sciences* XLII-5: 123–128.
https://doi.org/10.5194/isprs-archives-XLII-5-123-2018

Lhomme, Stéphane, Dong-Chen He, Christiane Weber, and D. Morin. 2009. "A new approach to building identification from very-high-spatial-resolution images." *International Journal of Remote Sensing* 30, is. 5: 1341–1354.
https://doi.org/10.1080/01431160802509017

Lv, Jingqin, and Jiangxiong Fang. 2018. "A Color Distance Model Based on Visual Recognition." *Mathematical Problems in Engineering* 2018: 4652526.
https://doi.org/10.1155/2018/4652526

Maarir, Abdelkrim, Belaid Bouikhalene, and Yassine Chajri. 2017. "Building Detection from Satellite Images based on Curvature Scale Space Method." *Walailak Journal of Science and Technology* 14, no. 6: 517–525.
https://doi.org/10.14456/vol14iss6pp%p

Maggiori, Emmanuel, Yuliya Tarabalka, Guillaume Charpiat, and Pierre Alliez. 2017. "Can Semantic Labeling Methods Generalize to Any City? The Inria Aerial Image Labeling Benchmark." *IEEE International Geo-science and Remote Sensing Symposium (IGARSS)*.
https://doi.org/10.1109/IGARSS.2017.8127684

Qin, Rongjun, and Wei Fang. 2014. "A Hierarchical Building Detection Method for Very High Resolution Remotely Sensed Images Combined with DSM Using Graph Cut Optimization." *Photogrammetric Engineering & Remote Sensing* 80, no. 9: 873–883.
https://doi.org/10.14358/PERS.80.9.873

Sharma, Gaurav, Wencheng Wu, and Edul N. Dalal. 2005. "The CIEDE2000 color-difference formula: implementation notes, supplementary test data, and mathematical observations." *Color Research & Application* 30, no.1: 21–30.
https://doi.org/10.1002/col.20070

Subudhi, Badri Narayan, Ishan Patwa, Ashish Ghosh, and Sung-Bae Cho. 2015. "Edge Preserving Region Growing for Aerial Color Image Segmentation." In *Intelligent Computing, Communication and Devices. Advances in Intelligent Systems and Computing*, edited by L. Jain, S. Patnaik, and N. Ichalkaranje, 309. New Delhi: Springer. https://doi.org/10.1007/978-81-322-2009-1_54

CHAPTER 3.3

PERSONALIZED DANGEROUS SITUATION DETECTION IN VEHICLE CABINS USING SMARTPHONE SENSORS

ALEXEY KASHEVNIK[1], ANDREW PONOMAREV[1], KSENIYA KARELSKAYA[1], MAKSIM REPP[2], ANASTASIIA CHERNYSHEVA[2], ANDREY KRASOV[3]

1 - SPIIRAS, St.Petersburg, Russia; ITMO University, St.Petersburg, Russia; alex-ey@iias.spb.su, ponomarev@iias.spb.su, k-karel@yandex.ru

2 - ITMO University, St.Petersburg, Russia; reppmg@gmail.com, chernysheva.anastasiia@gmail.com

3 - Bonch-Bruyevich St. Petersburg State University of Telecommunications, St.Petersburg, Russia; krasov@inbox.ru

Abstract. This paper proposes an approach for the detection of dangerous situations by a driver in a vehicle cabin using information from smartphone sensors. The approach presented utilizes thresholds that classify normal and dangerous states. A personalized algorithm is proposed with the aim of adapting threshold parameters according to the analysis of an individual driver. The authors propose a related workplace analysis for predicting and detecting dangerous situations through driver behaviour analysis and driver style identification, as well as approaches based on personalization and adaptation. Identification of dangerous states can be undertaken through analysis of information from smartphone sensors and a classification for this has been developed. This classification shows what information is needed for the identification of a dangerous situation. Then, the authors present algorithms for detecting drowsiness, distraction, aggressive driving, and high heat states. Following this, a personalization algorithm is proposed using the example of detecting the dangerous state of distraction.

1. Introduction

Traffic accidents remain a serious issue for many countries across the world. There has been a lot of research and development into the topic of autonomous driving and advanced driver assistance systems with the aim of reducing accidents on public roads. In the future, all vehicles on public roads are likely to be autonomous, but for the moment such a perspective appears fantastical. As such, future developments will involve approaches to detect dangerous situations with a driver in a vehicle cabin. Weidner, Transchel, and Weidner (2017) suggest that it is important for insurance companies to analyze driver profiles and offer discounts or surcharges on insurance premiums.

In recent years, many systems have been developed with the aim of monitoring drivers, recognizing dangerous states, and notifying them. The research and development community has proposed hardware systems integrated into vehicles and the use of special sensors and driver monitoring; as well as taking advantage of smartphone applications so as to use existing hardware that a driver will often own. The quality of these first solutions was reasonable, but such systems are usually only found in luxury vehicles and they are expensive for most drivers. The goal of the following chapter is to classify the dangerous states that can be detected in the vehicle cabin based on the monitoring the driver through built-in smartphone sensors and a camera mounted on the vehicle's windshield, as well as studying algorithms for these purposes.

This chapter extends a previous paper (Kashevnik, Karelskaya, & Repp, 2019), which considers the dangerous situation detection approach in the vehicle cabin. The chapter contains the following new issues:

- Related work analysis on the topic of personalization and adaptation for the detection of dangerous driving states has been carried out;
- The classification scheme for detected dangerous situations based on smartphone sensor information has been improved;
- An algorithm for adaptive threshold parameters has been developed.

The chapter is structured as follows. Section 2 describes related work in the area of dangerous situation detection through driver behavior analysis and driver style identification, as well as personalization and adaptation approaches. The dangerous situation classification, as well as algorithms for the detections of drowsiness, distraction, aggressive driving, and overheating are presented in section 3. This section also presents an algorithm for the use of adaptive threshold parameters for an individual driver. The main results are summarized in the conclusion.

2. Related Work

This section describes the analysis of dangerous situation detection through driver behavior analysis and driver style identification, as well as personalization and adaptation approaches aimed at learning based on analysis of a particular driver to improve the dangerous situation detection system.

2.1. Driver Behavior and Driver Style Analysis

Verwer et al. (2011) proposed an approach for studying the behavior of truck drivers (leading to excessive fuel consumption) based on recorded values from speed sensors, revolutions of the engine per minute, and fuel use. The authors proposed the RTI+ algorithm to detect this behavior. This approach was tested in the scenario of detecting excessive fuel consumption where the car accelerates too quickly. Speed, fuel, and engine sensors were used to analyze driver behavior.

Aoude et al. (2011) developed an algorithm for classifying the behavior of drivers at road intersections based on real data. The object of this study was the reaction of drivers to the prohibitory signal of the traffic light (red or yellow), as well as whether they stopped in front of the stop line, or left a distance of three meters. When a potentially dangerous driver behavior is recognized, the system should notify him/her about it. For dangerous behavior recognition, onboard vehicle sensors were used. For classification of drivers into "violation" and "docile", the support vector machine (SVM) and the hidden Markov model (HMM) approaches were used to model the basic patterns in the set of observations. A real-time data collection system was used to collect data at the intersection. This system includes four radar units to identify vehicles and measure vehicle speed, range, and lateral position; a GPS antenna; four video cameras; and a traffic analyzer to record the phase of a traffic light.

In Van Ly et al. (2013) the possibility of using vehicle inertial sensors from a CAN bus was studied. For these experiments, the authors used a Volkswagen Passat car. This was equipped with the appropriate sensors and a "vision system". The sensor package included: frontal radar, vehicle CAN signals, GPS, and front/rear camera(s). To study driver behavior, all types of maneuver were analyzed using three types of event: braking, acceleration, and turning. The SVM algorithm was used by the authors to analyze the dataset in the study.

Johnson et al. (2011) suggested that driving style can be divided into two categories: "typical" (non-aggressive) and aggressive. To increase driver awareness and safety, the authors proposed a system using dynamic time warping (DTW) and a smartphone with sensors (accelerometer, gyroscope, magnetometer, GPS, camera) for detecting, recognizing, and recording driver actions without external processing. The developed system is called "MIROAD" (mobile-sensor-platform) and is used for intelligent recognition of aggressive driving. The HMM and DTW algorithms were used by the authors to analyze the data.

In Khattak et al. (2018), the authors explored ways of reducing the effect of stress on driver behavior. The study assessed the impact of adaptive signal control technology (ASCT) on stress levels and driver behavior. The participants of the experiment drove along two roads, one of which was equipped with ASCT, while the other was not. During a trip, the system recorded data on the heart rate of the drivers and their behavior. The data were analyzed using correlation coefficients of analysis of variance. As a result, it was found that on the road equipped with ASCT, the heart rate of the drivers was significantly lower (about 10 beats), leading to a decrease in the latter's stress.

Sandeep et al. (2017) propose an original Internet of things module to protect people from deaths in traffic accidents caused by drunk driving. The proposed system uses the Raspberry Pi3 B (a single-board computer that is the size of a bank card). It includes a touch sensor, alcohol concentration detection sensor, face detection, and heart rate measurement. It also uses a GPS module and supports emergency notification, as well as automatic ignition shutdown.

In Dhar et al. (2014), the authors propose a highly efficient system to help a driver detect and give advance warning of dangerous vehicle maneuvers caused by aggressive driving or drunk driving. The system uses built-in smartphone sensors: an accelerometer and an orientation sensor. The mobile phone-based program calculates acceleration based on sensor readings and compares it to typical unsafe driving characteristics obtained from actual driving tests.

Wu et al. (2016) propose a new driving behavior recognition system based on a particular physical model and using motion sensing data to improve traffic safety. The system uses a nine-axis motion sensor that includes a three-axis accelerometer, a three-axis gyroscope, and a three-axis magnetometer. To extract the data, the authors used a Kalman filter and an adaptive time-window. Based on the extraction function, various classifiers are performed to recognize different types of driving. The system classifies normal driving behavior and aggressive driving behavior based on characteristics such as acceleration, braking, changing lanes, and turning, with an accuracy of 93.25 %.

Kumar & Patra (2018) developed a system for monitoring driver drowsiness in real time. The system uses a smartphone camera to record video of a driver's face. The system fixes facial landmarks on the detected face and subsequently calculates the proportions of the eyes, the ratio of mouth opening, and the length of nose ratio. Depending on the calculated values the system detects a state of drowsiness based on selected adaptive threshold values. The SVM and histogram of oriented gradient (HOG) algorithms were used by the authors to analyze the data.

Chen et al. (2015) distinguish six types of abnormal driver behavior: weaving; swerving; side slipping; fast U-turn; turning with a wide radius; and sudden braking. The authors propose a system for discovering each type of dangerous movement using built-in smartphone sensors: a 3-axis accelerometer; a 3-axis orientation sensor; and a digital video recorder. To identify behavioral patterns, an SVM algorithm is used.

Kalra and Bansal (2014) implemented a review of existing systems for the detection of states and driver behavior with the help of smartphone sensors. They list commonly used sensors, including: accelerometer; gyroscope; GPS magnetometer; camera; and microphone. Some algorithms that help analyze data are also presented, including: DTW; SVM; machine learning algorithms; artificial neural networks; decision trees; and HMM.

Abulkhair et al. (2015) propose a driver fatigue detection system (DFS), which monitors the driver and gives a warning signal if the driver falls asleep at the wheel. The system runs on a smartphone. The camera is used to detect the following states: yawning; nodding of the head; and movement of the eyelids. To analyze the data, the authors used Haar-like classification cascades and a pattern matching algorithm.

Galarza et al. (2018) propose a system to detect driver drowsiness in real time. This system is a smartphone application in Android OS and uses a camera to detect the driver's face. The accuracy of detecting drowsiness in this system is estimated at 93.37 %. The system works adequately in natural light and regardless of whether the driver wears accessories, such as glasses, headphones, or a cap.

In the scope of this paper's analysis, the following characteristics are identified. The correspondence of the considered systems and relevant characteristics is presented in Table 1. In the table, "+" means that the system shows a satisfactory correspondence; "–" means an unsatisfactory correspondence; and "0" suggests that it was not possible to recognize a correspondence. These characteristics include:

- Aggressive driving detection
- Fatigue detection
- Drowsiness detection
- Stress level detection
- Drunken driving detection
- Use of magnetometer
- Use of accelerometer
- Use of orientation sensor
- Use of camera
- Use of GPS
- Use of speed sensor
- Use of gyroscope
- Pulse measurement
- Use of microphone
- SVM algorithm
- HMM algorithm
- DTW algorithm
- HOG algorithm
- Haar-like classification cascades
- Pattern matching algorithm

Table 1. Comparison of the considered systems

Article /comparison criteria	Aggressive driving	Fatigue detection	Drowsiness detection	Stress level detection	Drunken driving	Use of magnetometer	Use of accelerometer	Use of orientation sensor	Use of camera	Use of GPS	Use of speed sensor	Use of gyroscope	pulse measurement	Use of microphone	SVM algorithm	HMM algorithm	DTW algorithm	HOG algorithm	Haar-like classification	Pattern matching algorithm
(Verwer et al., 2011)	+	-	-	-	-	-	-	-	-	-	+	-	-	-	-	-	-	-	-	-
(Aoude et al., 2011)	+	-	-	-	-	-	-	-	+	+	-	-	-	-	+	+	-	-	-	-
(Van Ly et al., 2013)	+	-	-	-	-	-	-	-	+	+	-	-	-	-	+	-	-	-	-	-
(Johnson & Trivedi, 2011)	+	-	-	-	-	-	+	-	-	+	-	+	-	-	-	+	+	-	-	-
(Khattak et al., 2018)	0	0	0	0	0	-	-	-	-	-	-	-	+	-	-	-	-	-	-	-
(Sandeep et al., 2017)	-	-	-	+	-	-	-	-	-	-	-	-	+	-	-	-	-	-	-	-
(Dhar et al., 2014)	+	-	-	-	-	-	+	+	-	-	-	-	-	-	-	-	+	-	-	-
(Wu et al., 2016)	+	-	-	-	-	+	+	+	-	-	-	-	-	-	+	-	-	-	-	-
(Kumar & Patra, 2018)	-	-	+	-	-	-	-	-	+	-	-	-	-	-	+	-	-	+	-	-
(Chen et al., 2015)	+	-	-	-	-	-	+	+	-	-	-	-	-	-	+	-	-	-	-	-
(Kalra & Bansal, 2014)	+	-	-	-	-	+	+	-	+	+	-	+	-	+	+	+	+	-	-	-
(Abulkhair et al., 2015)	-	+	-	-	-	-	-	-	+	-	-	-	-	-	-	-	-	-	+	+
(Galarza et al., 2018)	-	-	+	-	-	-	-	-	+	-	-	-	-	-	-	-	-	-	-	-

2.2 Personalization and Adaptation

There is a plenty of evidence, both common-sense and scientific (Hasenjager & Wersing, 2018), that drivers vary significantly in terms of their goals (e.g., private or commercial) and driving habits. For some typical driver assistance functions this difference does not really matter; for other functions (especially for functions based on the processing of visual/audio signals/commands from the driver or recommending actions to be taken) the utilization of personal characteristics and preferences may improve the quality of the system. Besides, if a driver's expectations about the functionality provided by the assistance system are not met, it may decrease trust in the system and lead to it being turned off, therefore ignoring its potential utility and positive influence on safety (Butakov & Ioannou, 2015; Butakov & Ioannou, 2016; Dang & Johannes, 2018).

The particular characteristics and values that are personalized depend on the kind of driver support. Characteristics that are used for monitoring the driving state can be divided into three categories:

- *Driving vehicle characteristics*. Every vehicle model has its own characteristics, including structure, weight, and performance, according to which the vehicle will behave differently on the road. An application that calibrates (and adapts) itself to detect events with respect to a particular vehicle is SenseFleet (Castignani et al., 2015). Experimental evaluation of the proposed approach has shown that different car models may display different "jerk" deviation. Other parameters that are monitored in acceleration and steering event detection are average yaw rate, speed variation, and bearing variation.
- *Physiological driver characteristics*. A drowsy state can be recognized by the eyes closing, the rate of eye-blinking, yawning, and dropping of the head. Kumar and Patra (2018) propose drowsiness detection by three feature: eye aspect ratio (EAR: indicating eye blinking); mouth opening ratio (MOR: indicating yawning); and nose length ratio (LNR: indicating nodding of the head, the ratio of nose length to average nose length is a measure of the head dropping down). All these features are calculated through the extraction of facial landmarks from a driver's image using image recognition methods. Eye size (mouth and nose) can differ from person to person, so it is necessary to have an adaptation phase. During this phase, the average values of eye and mouth size and nose length for every person are calculated. These values are then used to evaluate the state of alertness. Deviation from these values can be used as an indicator of a drowsy state. Choi and Kim (2017) pro-

pose a personalized classifier for the detection of driver drowsiness based on physiological data. It is based on a set of driver characteristics measured through a wearable device, including: photoplethysmogram (PPG); galvanic skin response (GSR); temperature (TEM); acceleration; and rate of rotation.

- *Driver behavior characteristics*. Drowsiness can also be detected by deviations in driving style and corresponding indicators (e.g., lane deviation indicators: time to line crossing, mean squared error for lane position, and heading error). These lane deviation indicators proceed on the assumption that a drowsy driver has problems in maintaining lane discipline and can be used as features of a model maximizing detection accuracy (Daza et al., 2014). Wang et al. (2016) propose forward collision detection where a driver is described in terms of their risk perception (an unobservable variable) of a situation, which is assumed to depend on two observable variables: headway time and inverse time-to-collision. An important area of behavioral characteristics focuses on a temporal model of a driver, connecting the driving situation (application-specific), the state (keeping in lane, lane changes etc.), and control inputs applied in such a situation and state (steering, acceleration). For example, Lefevre et al. (2015) used such a model to detect possible lane departure.

There are basically four types of methods for adapting a driver assistance system to the behavior of a particular driver (Ponomarev & Chernysheva, 2019), including: adaptive thresholds; classifiers; personalized function dependencies; and personalized behavior patterns. Using an adaptive threshold, which separates the acceptable values of some observable parameter from the unacceptable ones, is a popular technique in the implementation of driver assistance. This technique has been used in various forms of driver monitoring (Castignani et al., 2015; Arroyo et al., 2016; Kumar & Patra, 2018) and maneuver assistance systems (Butakov & Ioannou, 2015). Basically, there are three ways to detect the personalized value of a threshold:

- based on the average for a particular driver/vehicle;
- by the maximum likelihood method, under the assumption of some specific distribution of the respective characteristic (a more elaborated example of the previous one, as averaging can be seen as the maximum likelihood estimation under the assumption of a normal distribution);
- fuzzy logic.

In a more general case, a decision rule for separating acceptable states from non-acceptable ones can be non-trivial and has to be modeled according to a set of non-observable parameters of some driver-specific classification model. Personalization in this case is usually done by creating a specific classification model and learning of the parameters by some machine learning algorithm (this approach is taken in papers such as Dang & Johannes, 2018; Wang et al., 2016; Choi & Kim, 2017). Some authors propose a model of personalized functional dependencies between some parameters. In some cases, such dependencies play a role very similar to that of a classifier (e.g., in Butakov & Ioannou, 2015) and a linear regression is used to predict an acceptable gap for a lane change according to the current road situation. In some driver assistance systems, driver-specific temporal sequences (e.g., steering sequences) are learned. These sequences are then exploited to predict the future state of the car and recognize the driver's intent. One of the most popular learning frameworks for this kind of driver model uses the hidden Markov model (HMM). For example, Lefevre et al. (2015) used HMM to connect the driving situation, the state, and the control inputs from the driver to detect possible lane departures.

One of the most important problems in adaptation to a particular driver is that the sample of driver behavior is very limited (Daza et al., 2014); this is why personalization models with few parameters have become very popular. The approach proposed in this paper also uses an adaptation technique with very few parameters (to be specific, it makes use of an adaptive threshold).

3. Dangerous Situation Detection

This section discusses the proposed dangerous situation detection approach. It considers an identified set of dangerous in-cabin situations that can be detected using information from smartphone sensors. The smartphone is located inside the vehicle cabin. The front-facing camera of the smartphone tracks the driver's face. Dangerous situation detection algorithms are considered and the adaptiveness of parameters that allow tuning of the algorithms to an individual driver are discussed.

3.1 Classification of Dangerous Situation Detection based on Information from Smartphone Sensors

The related work review allows us to highlight the parameters that are needed for dangerous situation detection algorithms and are accessible to built-in smartphone sensors (see Table 2). Based on these parameters the detection of the following dangerous states has been identified:

- Drowsiness
- Distraction
- Stress Level
- Aggressive Driving
- Drunk Driving

It has been proposed that these dangerous states should be classified as online, semi-online, and offline (see Fig. 1). Online dangerous states have to be detected in real time by the smartphone while driving. The detection time depends on the context situation (driver reaction time, vehicle speed etc.), but it is around two seconds. These dangerous states include drowsiness and distraction. If the driver is drowsy or distracted, the system should influence him/her in real time to avoid an accident.

Semi-online dangerous states are dangerous states that can be detected using a process that takes more than two seconds, but that should be detected by smartphone as rapidly as possible. Such factors include aggressive driving and a rapid pulse, which can be predictive of danger.

Table 2. Accessible parameter classification for dangerous situation detection

Parameter	Sensor	Description
Eye opening ratio	Camera	Detect drowsiness state by eyes
PERCLOS		
Mouth opening		Detect drowsiness state by mouth
Tilt angle of head		Detect drowsiness state by head
Rotation angle of head		Detect distraction state by head
Pulse		Detect driver pulse and stress level
Acceleration	Accelerometer	Detect aggressive driving state by vehicle acceleration
Vehicle turn	Gyroscope	Detect aggressive driving state by vehicle turning
Voice	Microphone	Detect drowsiness state
Light level	RGB-light sensor	Detect light level that is influenced to the dangerous state detection using smartphone camera
Geolocation	GPS / GLONASS	Give a recommendation to driver where he can sleep id he falls asleep

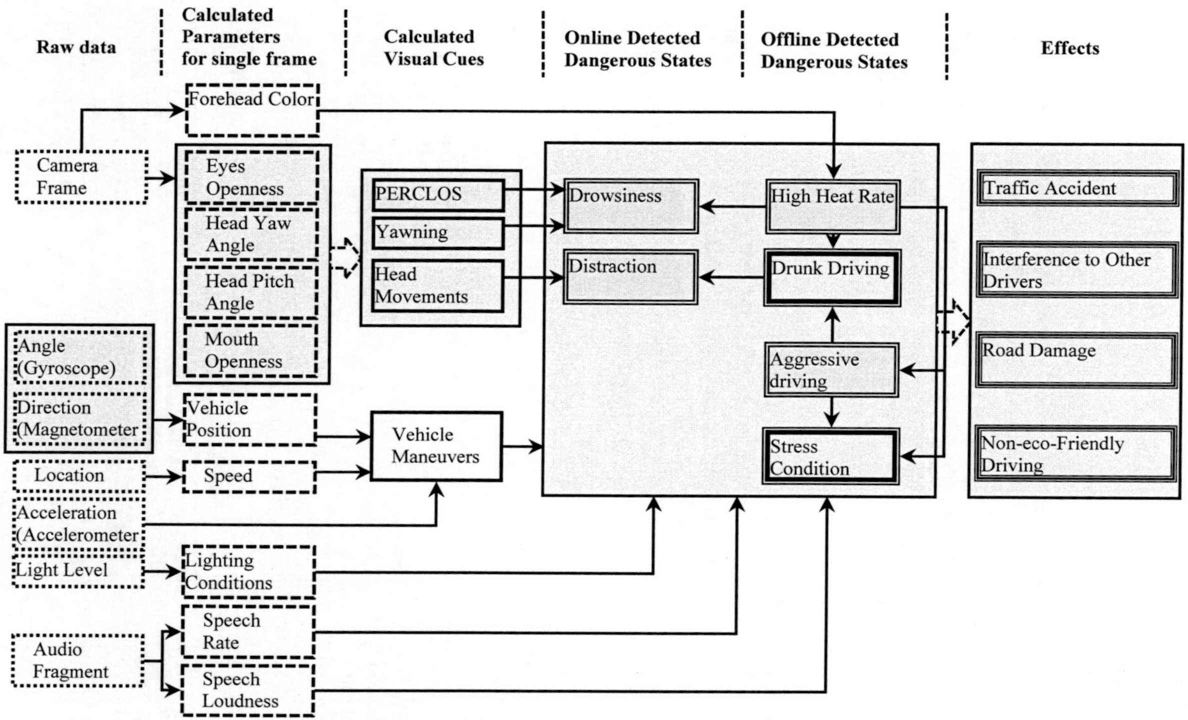

Fig. 1. Classification of dangerous situation detection using smartphone sensors

Offline dangerous states are dangerous states that are detected in the cloud (not on the smartphone). The detection time can be greater than a minute since the calculations have to be implemented and a lot of data has to be analyzed. These situations are detected on a basis not limited to data collected in real time, but data previously collected and stored. These dangerous state factors include: drunk driving and stress level.

The following effects have been detected as a consequences of dangerous states: traffic accidents; impediments to other road users; road damage; and non-eco-friendly driving.

Drowsiness is a state that drivers of long-distance buses, those driving at night, and truck drivers are susceptible to. Characteristic features of drowsiness include: yawning, nodding of the head, closing of the eyes, and blinking. To recognize these states a smartphone camera is used.

Distraction is a condition characterized by a driver turning their head to the side in such a way that they are not able to monitor the condition of the road at a point in time.

A high pulse rate is a state characterized by rapid heartbeat caused by the external environment of the driver. This state allows us to detect a drunk driving state. A smartphone camera can be used to recognize an increased heart rate.

The stress level state of the driver sees the external environment have a negative impact on their psycho-emotional state. This state can cause aggressive driving. To detect the level of stress, a gyroscope, and accelerometer sensors can be used.

Drunk driving is a state in which a driver's blood contains a certain concentration of alcohol. The consequence of this condition is the manifestation of various types of aggressive driving. To detect drunk driving, the following sensors are used: camera; gyroscope; accelerometer; and GPS/GLONASS.

Aggressive driving is characterized by weaving, swerving, side slipping, fast U-turns, turning with a wide radius, and sudden braking. The following sensors are used to recognize this state: accelerometer; gyroscope; GPS; orientation sensor; magnetometer; and camera.

3.2 Dangerous Situation Detection Algorithms

Algorithms for online and semi-online detection of dangerous states based on smartphone sensors are discussed in the following: drowsiness; distraction; aggressive driving; and a rapid pulse rate. In the future, the authors plan to develop algorithms for drunk driving and stress level detection.

3.3 Drowsiness

According to the literature, drowsiness can be detected by analysis of the following three parameters: the ratio of eye opening (*E*); the ratio of mouth opening (*M*); and the angle of the head (*NA*). Below, an algorithm is proposed based on analysis of the relevant work. It consists of the following two steps: parameter calibration and drowsiness detection.

In the first step, a number of frames are analyzed (some authors propose 300 frames). Based on these frames, the parameters *E*, *M*, and *NA* are detected and the threshold values are calculated: E_0, M_0, and NA_0. In the second step, the measurement of *E*, *M*, and *NA* is implemented. During the real-time measurement of the appropriate indicators, the state of drowsiness is based on the threshold values of E_0, M_0, and NA_0.

The formula for calculating the coefficients *E* and *E1* for the left eye is presented below; the coefficient *E* for the right eye is similarly calculated.

$$E = \frac{(p2-p6)+(p3-p5)}{2(p4-p1)},$$

where [p_1; p_6], [p_7; p_{10}] are points on the left and right eyes necessary for the calculation of the eye opening ratio (see Fig. 2).

Fig. 2. Points detected by smartphone camera for monitoring a driver's eyes

PERCLOS is a measure of drowsiness detection, being the percentage closure of the eyelid over the pupil over time and reflects the slow closure of the eyelid, while not blinking. The definition of the formula is presented below, where N_m is the total number of frames taken per minute and N_a is the number of frames during which the eyes are open. Relevant work suggests that if the PERCLOS parameter is greater than 70 % (parameter *D* in the algorithm scheme), then a state of drowsiness is discovered. The algorithm for detecting drowsiness by monitoring of the eyes is shown in Fig. 3.

Driver drowsiness can also be detected by the degree of mouth opening (yawning). Yawning can be defined as follows: if the mouth opening ratio M_1 increases monotonically and then tends towards the value M_0, this state is classified as yawning and, therefore, drowsiness.

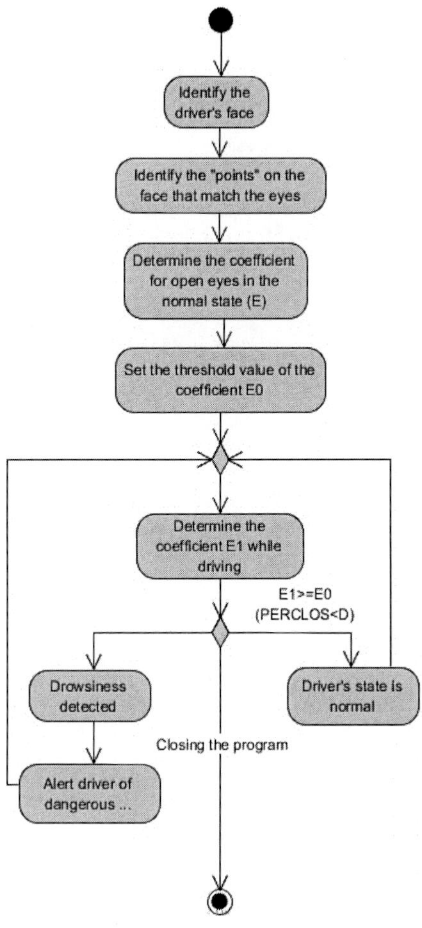

Fig. 3. Drowsiness detection algorithm using eye monitoring

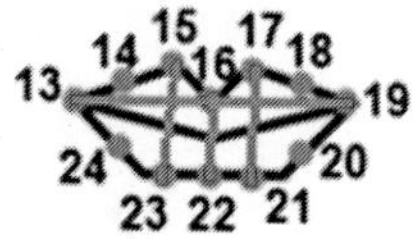

Fig. 4. Points detected by smartphone camera for monitoring a driver's mouth

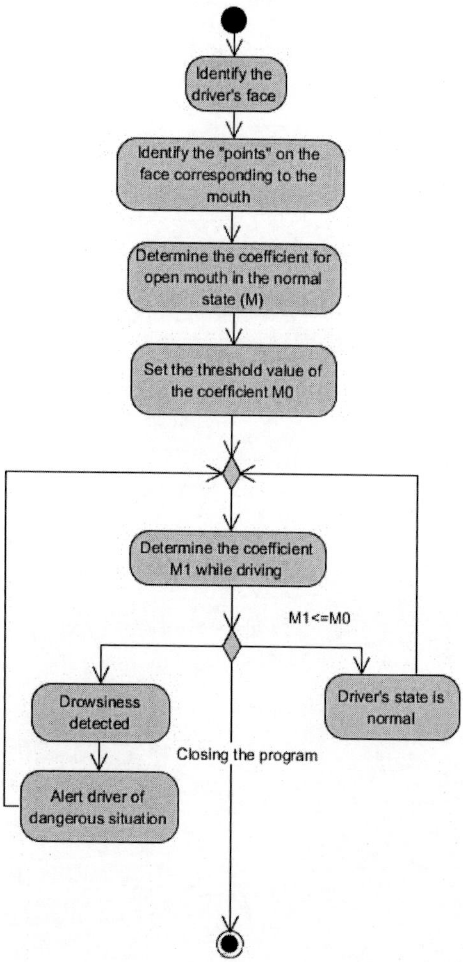

Fig. 5. Drowsiness detection algorithm by monitoring the mouth

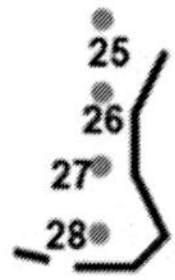

Fig. 6. Points detected by smartphone camera for monitoring a driver's nose

The formula for calculating the coefficient *M* is presented below

$$M = \frac{(p15 - p23) + (p16 - p22) + (p17 - p21)}{2(p19 - p13)}$$

$$PERCLOS = \frac{N_m - N_a}{N_m} * 100\%$$

where [p13; p24] are points on the mouth necessary for calculating the mouth opening ratio (see Fig. 4). An algorithm for the detection of drowsiness by mouth monitoring is shown in Fig. 5.

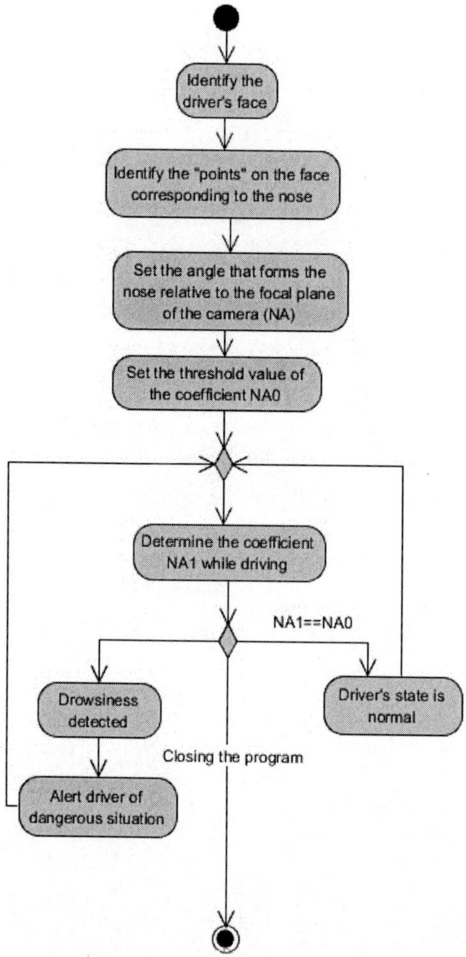

Fig. 7. Drowsiness detection algorithm through monitoring of the tilt angle of the nose

3.4 Distraction

The dangerous state of distraction is usually detected using measurement of the left/right head angle. Initially, the head angle is calibrated with the situation of drowsiness detection. The angle α^{yaw} is the calibrated offset. The distraction dangerous state is detected when driver turns their head more than an angle of $\alpha^{yaw}{}_{cr}$. An algorithm for distraction detection by monitoring the angle of rotation of the nose is shown in Fig. 8.

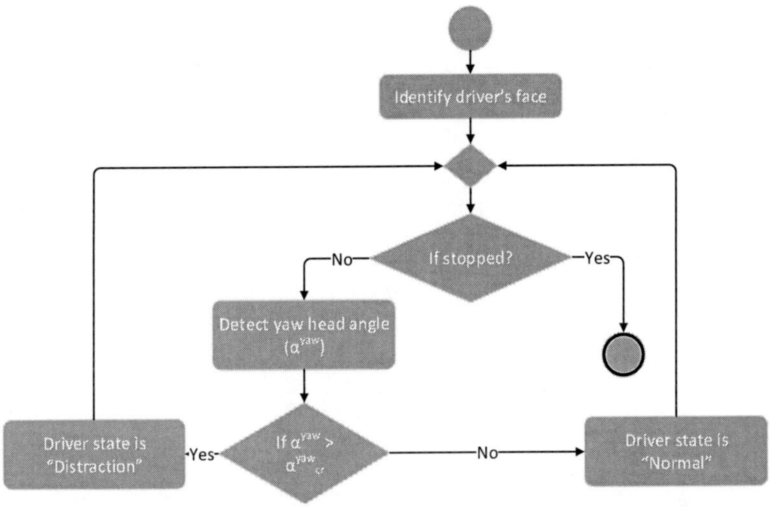

Fig. 8. Distraction detection algorithm using rotation angle of nose monitoring

The state of distraction can also be detected by a driver's nodding head. The driver's head can either bend down or lean back. Firstly, the driver's nose angle has to be calibrated, and then, if the driver begins to fall asleep, the angle changes up or down, revealing a state of drowsiness.

The formula for calculating the coefficient *NA* is presented below

$$NA = \frac{nose\ length(p28 - p25)}{average\ nose\ length}$$

where [p25; p28] are points on the nose used to calculate its tilt angle (see Fig. 6). An algorithm for the detection of drowsiness through monitoring of the tilt angle of the nose is shown in Fig. 7.

3.5 Aggressive Driving

The state of aggressive driving is detected using accelerometer and orientation sensor values, classified into six types of behavior: weaving, swerving, side slipping, fast U-turns, turning with a wide radius, and sudden braking.

The algorithm for the detection of aggressive driving by monitoring vehicle coordinates is shown in Fig. 9.

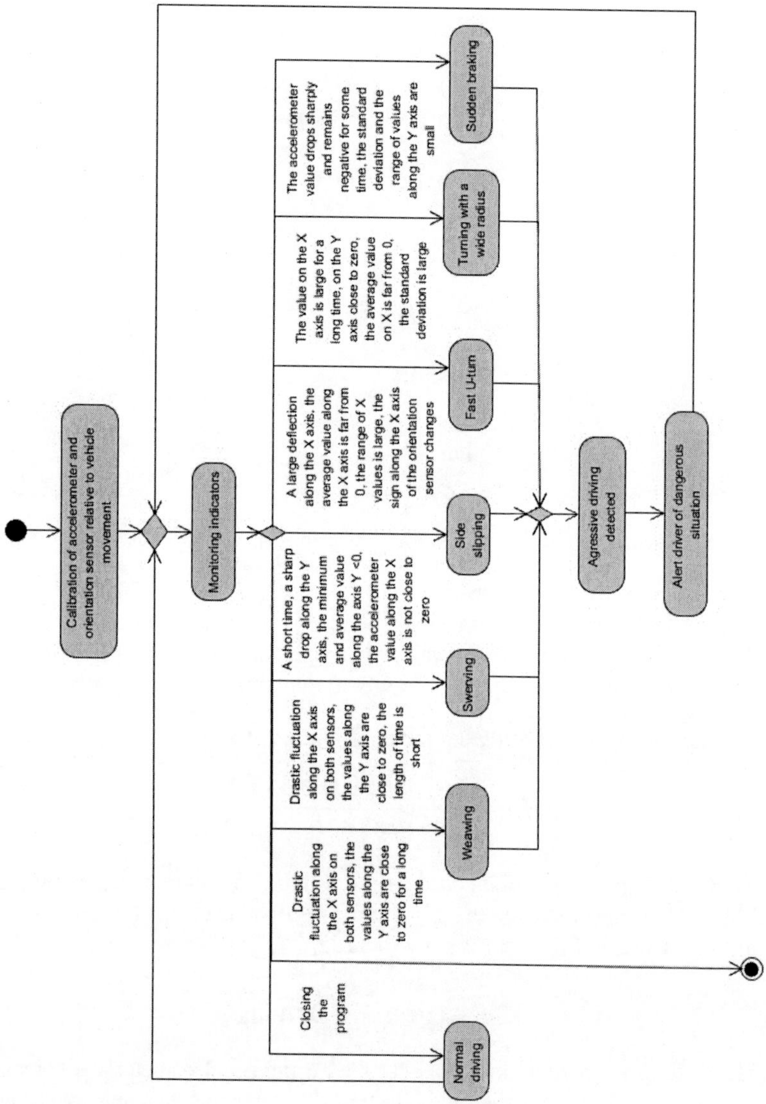

Fig. 9. Aggressive driving detection algorithm by monitoring vehicle movement

3.6 Heat Rate Detection

Research in the area of pulse detection by camera shows that the primary scheme is as follows. The face of the person is tracked, then the area for analysis is determined, the mean color is calculated and transformed to the frequency characterizing the pulse. Fig. 13 shows two different algorithms for pulse detection implemented in the following applications: webcam-pulse-detector (https://github.com/thearn/webcam-pulse-detector); and motion-pulse (https://github.com/COMP6206-vision-based-pulse-detection/python-motion-pulse).

Both algorithms have the following structure: data extraction; artificial sample rate increase; frequency extraction; remove abnormalities; and filter values that are unrelated to the pulse steps.

Webcam-pulse-detector algorithm relies on color change in a region of a user's face at step 1. A user's region of interest is shown in Fig. 10 by a green rectangle. To extract regions of interest from a facial image, the image should be recognized utilizing OpenCV, and the forehead region has to be cut from the facial image for each frame in a video. The region of interests is a set of pixels that have values for red, green, and blue components in the form of an integer value in the range [0, 255]. For each component, the algorithm calculates its mean value by averaging the component value for all pixels in the region of interest. The result is three mean values for the red, green, and blue components. The mean value of those three numbers is the data extracted from the frame and is added to the data buffer (Verkruysse, Svaasand & Nelson, 2008) (see example in Fig. 11).

The motion-pulse algorithm uses facial point movements to estimate the pulse. The algorithm selects a series of points on the face using OpenCV (Fig. 12) and then utilizes the feature tracking algorithm to track point coordinates. Point coordinates for every frame of the video are extracted (Balakrishnan, Durand, & Guttag, 2013).

At step 2, the interpolation of the results achieved in step 1 is implemented. Since the smartphone camera has a frame rate of 30 frames per second, but electrocardiography devices record at a sample rate of 250 Hz, these values need to be synchronized. The motion-pulse algorithm uses cubic-spline interpolation to 250 Hz. Since the rate of frames per second is not constant, the webcam-pulse-detector algorithm uses one-dimensional linear interpolation to make the intervals between the frames even.

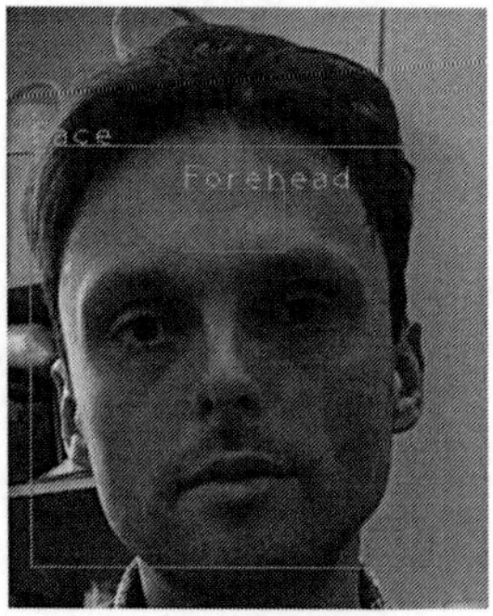

Fig. 10. A person's face and region of interest highlighted by a rectangle

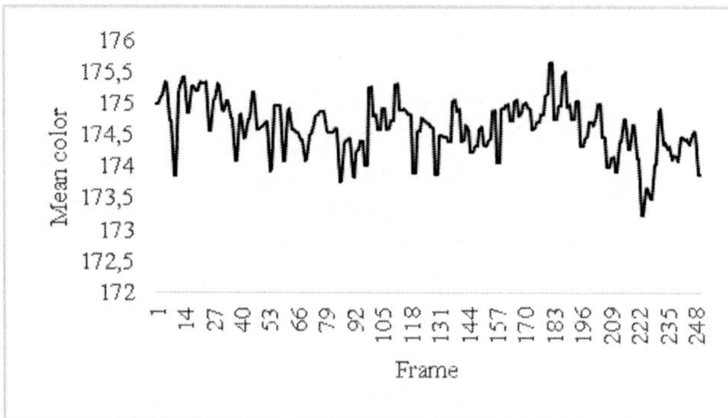

Fig. 11. Example of mean color values from the region of interest

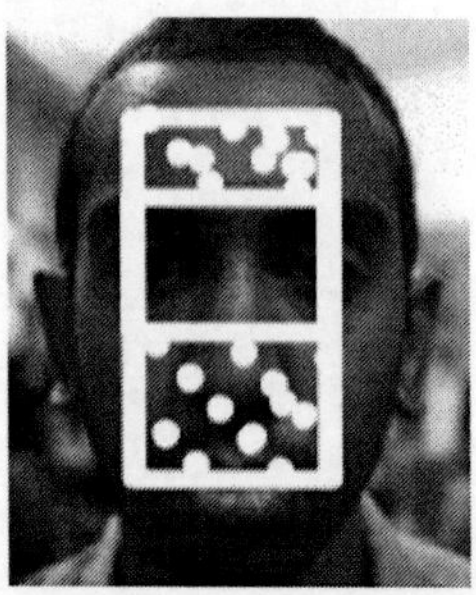

Fig. 12. Points on a person's face tracked by the motion-pulse algorithm

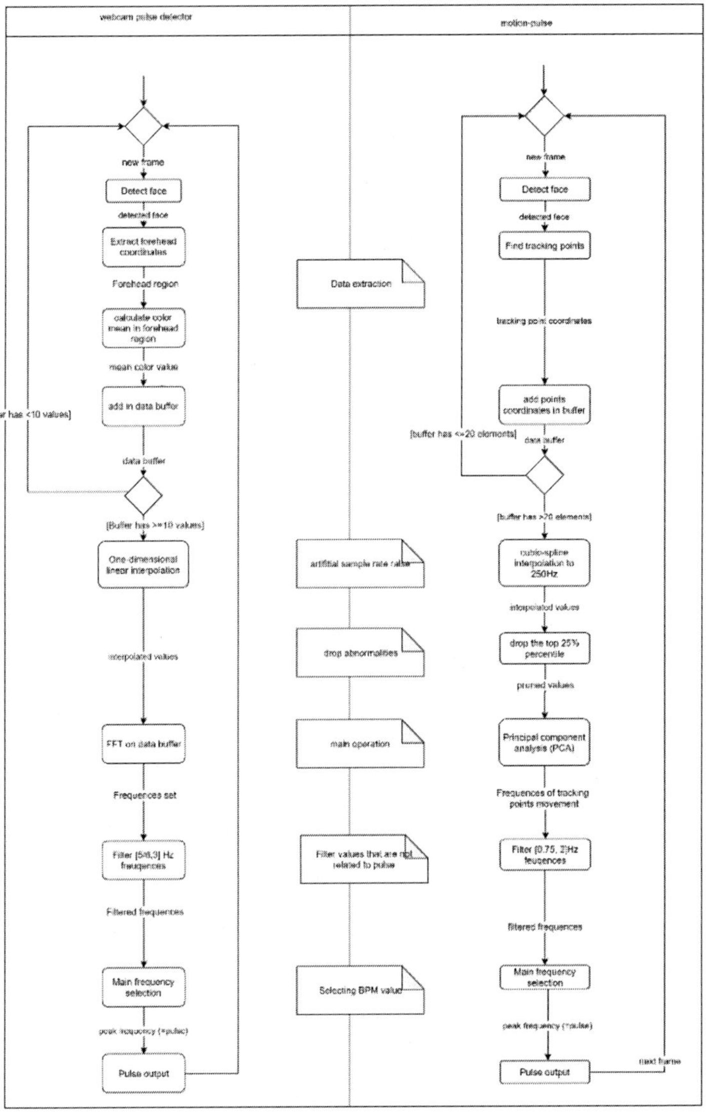

Fig. 13. Pulse detection algorithms based on information from front-facing smartphone camera measurements

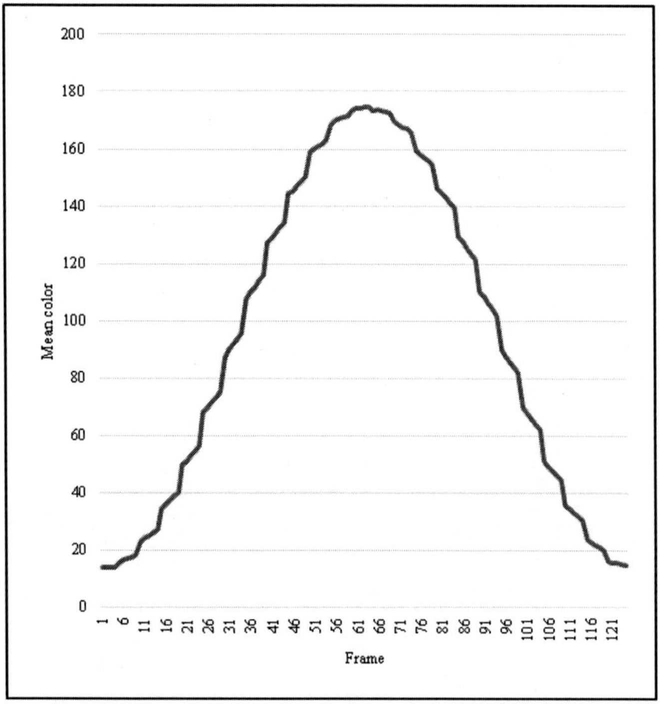

Fig.14. Example of the data with a Hamming window applied

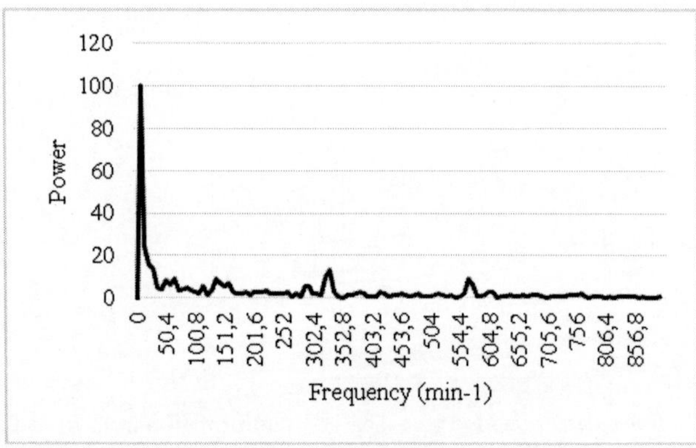

Fig. 15. Example of fast Fourier transforms

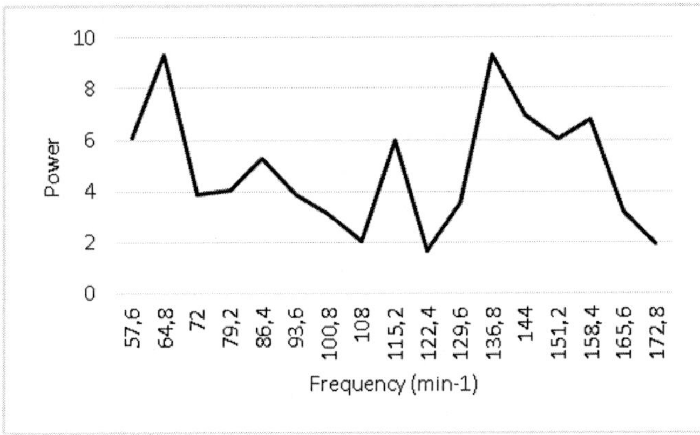

Fig. 16. Filtered data between 50 and 180 beats per minute

The 3rd step aims to remove abnormalities when analyzing the data. This is important in applying the motion-pulse algorithm. Recorded motions are not just pulse motions, but also include swallowing and adjustments in posture etc. To deal with these, 25 % of movements with the highest amplitudes are dropped.

The goal of step 4 is to convert the interpolated color data to a frequency. Before this conversion, a Hamming window function is applied (see Fig. 14).

To implement this conversion, the webcam-pulse-detector uses fast Fourier transforms (Verkruysse, Svaasand & Nelson, 2008) on interpolated data to detect power and phase spectra (see Fig. 15). After transformation, frequencies should be filtered; the result is the powers of all frequencies between 50 and 180 beats per minute (see Fig. 16). The highest obtained frequencies are considered to offer a suitable pulse estimation. The motion-pulse algorithm uses principal component analysis to select the component whose temporal power spectrum best matches a pulse using fast Fourier transformation (in the case considered, this is 136.8 bpm).

At step 5, the achieved frequencies are filtered. Pulse values have certain limits and values outside those limits should be filtered out. In the webcam-pulse-detector the limits are [50, 180] beats per minute; in the motion-pulse algorithm the limits are [45, 120] beats per minute.

In the final step the resultant pulse value is selected. From the filtered values, the pulse estimation is the frequency with the highest power.

3.7 Algorithm of threshold parameters adapted for a particular driver

The proposed personalization algorithm focuses on the adjustment of the algorithms for detecting dangerous situations to the individual characteristics of the driver (psychological, physiological, and behavioral). As was already noted earlier, adaptation to the driver, on the one hand, improves the quality of the system's recommendations and, on the other hand, increases the acceptance of the system and the driver's confidence in it.

This personalization is currently done using adaptive thresholds. This is for several reasons. First, a significant problem in the personalization of support systems is the limited sample that one driver provides, which makes it difficult to apply personalization methods that are characterized by a large number of parameters. Second, methods for detecting adaptive thresholds are characterized by their simplicity. Finally, in some cases, despite this simplicity, they allow the significant improvement of system quality and the degree of trust in it. All these factors detect the choice of personalization based on adaptive thresholds at this stage of the project.

Let us consider the application of adaptive thresholds for improving detection of driver distraction based on the driver's head rotation. It is assumed that the driver's attention has weakened (he/she is distracted) if the yaw angle of the driver's head α^{yaw} to the direction of movement on several consecutive frames recorded by the camera of the smartphone turns out to be greater in terms of absolute value than some critical value $\alpha^{yaw}{}_{cr}$. This critical value depends on the driver's habits and the characteristics of the road segment and an adaptive threshold is estimated according to the driver's history and current situation.

Threshold estimation includes the following steps:

1) Collect driving history (GPS data, head position, etc.) and transfer it to the cloud service, where all registered events are stored and grouped by driver. Analytical algorithms thus have the opportunity to access a "record" of driving parameters on any road segment.

2) From the data for driver *d* (for whom the threshold is being estimated) we select segments with a speed greater than a set threshold (e.g., 10 km/h).

3) The data is split into segments of straight movement and left and right turns. The split is done using GPS data. First of all, segments of straight movement are identified. The criterion for straight movement is that the difference between the azimuths within the segment does not exceed a certain angle (15°). These segments are classified as straight, while the rest of the segments are considered as turns; the direction of a turn is detected by the angle between the straight segments adjacent to it. Each recorded track, initially containing GPS readings, the orientation of the driver's head, the speed, and other parameters at each point in time, is complemented by a label describing the nature of movement at that particular moment: *Forward*, *Left*, *Right*.

4) A ranked sample of the absolute value of the driver's head yaw angle $|\alpha^{yaw}|$ is constructed for each segment of each movement type ($t \in$ {*Forward*, *Left*, *Right*}). Given the type of movement, the threshold $\alpha^{yaw}{}_{cr}(d,t)$ is set to be equal to the ninth decile of the constructed ranked sample.

5) The values of the thresholds are transferred to the driver's smartphone and are used when detecting dangerous driver situations.

Steps 1-5 are periodically repeated.

The algorithm is graphically visualized in the UML activity diagram in Fig. 17.

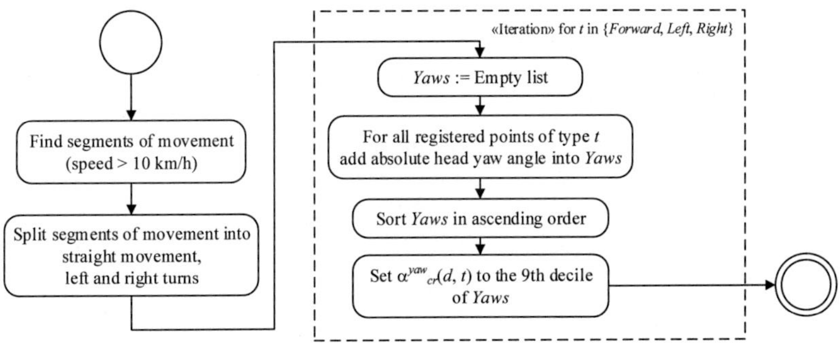

Fig.17. Algorithm of threshold parameter adaptation

4. Conclusions

This chapter discusses relevant work on the topic of dangerous situation detection through driver behavior analysis, as well as personalization and adaptation approaches. This enables the classification of dangerous states that can be detected in the vehicle cabin using a smartphone camera and sensors with a list of parameters tracked by mobile phone. Dangerous situations are classified as detected "online", which the smartphone can process without additional processing resources, and those detected "offline" are processed in the cloud. Four algorithms are considered: drowsiness; distraction; aggressive driving; and pulse detection. The dangerous states of drowsiness and distraction were implemented in the DriveSafely[1] application developed by authors, which is accessible through Google Play. The algorithm for detection of the threshold parameters provides possibilities for the analysis of driver behavior and adaptation to the dangerous situation detection algorithms based on this behavior. It should be noted that increased pulse detection is important not only for the vehicle driver, but also for e-bikes (Kashevnik et al., 2018).

Acknowledgment

The research presented in this paper was undertaken at ITMO University (funded by the Russian Science Foundation (project # 18-71-10065). Work on the topic of personalization and adaptation has been done in SPIIRAS within the scope of Russian State Research #0073-2019-0005. Experiments with heat rate recognition were conducted in Bonch-Bruyevich St. Petersburg State University of Telecommunications.

References

Abulkhair, M., Alsahli, A., Taleb, K., Bahran, A., Alzahrani, F., Alzahrani, H., Ibrahim, L. (2015). Mobile Platform Detect and Alerts System for Driver Fatigue, *Proceedings of the 2015 International Conference on Soft Computing and Software Engineering (SCSE'15)*, Vol. 62, pp. 555–564.

Aoude, G., Desaraju, V., Stephens, L., How, J. (2011). Behavior Classification Algorithms at Intersections and Validation using Naturalistic Data, *2011 IEEE Intelligent Vehicles Symposium (IV)*, pp. 601–606.

Arroyo, C., Bergasa, L.M., & Romera, E. (2016). Adaptive Fuzzy Classifier to detect Driving Events from the Inertial Sensors of a Smartphone. *IEEE 19th International Conference on Intelligent Transportation Systems (ITSC)*.

---

[1] https://play.google.com/store/apps/details?id=ru.igla.drivesafely&hl=ru

Balakrishnan, G., Durand, F., Guttag, J. (2013). Detecting Pulse from Head Motions in Video, *IEEE Conference on Computer Vision and Pattern Recognition*. pp. 3430–3437.

Butakov, V., & Ioannou, P. (2015). Personalized Driver/Vehicle Lane Change Models for ADAS. *IEEE Trans. Veh. Technol.*, vol. 64, № 10, 4422–4431.

Butakov, V.A., & Ioannou, P. (2016). Personalized Driver Assistance for Signalized Intersections Using V2I Communication. *IEEE Trans. Intell. Transp. Syst.*, vol. 17, № 7, 1910–1919.

Castignani, G., Derrmann, T., Frank, R., & Engel, T. (2015). Driver Behavior Profiling Using Smartphones: A Low-Cost Platform for Driver Monitoring. *IEEE Intelligent Transportation Systems Magazine*, vol. 7, issue 1, 91–102.

Chen, Z., Yu, J., Zhu, Y., Chen, Y., Li, M. (2015). Abnormal Driving Behaviors Detection and Identification Using Smartphone Sensors. *12th Annual IEEE International Conference on Sensing, Communication, and Networking (SECON)*, pp. 524–532.

Choi, M., & Kim, S.W. (2017). Online SVM-based personalizing method for the drowsiness detection of drivers. *Proc. Annu. Int. Conf. IEEE Eng. Med. Biol. Soc. EMBS*, 4195–4198.

Dang, H. & Johannes, F. (2018). Exploiting Maneuver Dependency for Personalization of a Driver Model Approach on Personalization. *Conference "Learning, Knowledge, Data, Analytics", LWDA 2018, CEUR Workshop Proceedings*, vol. 2191, 93–97.

Daza, I., Bergasa, L., Bronte, S., Yebes, J., Almazan J., & Arroyo, R. (2014). Fusion of Optimized Indicators from Advanced Driver Assistance Systems (ADAS) for Driver Drowsiness Detection. *Sensors (Basel)*, 14(1), 1106–1131.

Dhar, P., Shinde, S., Jadav, N., Bhaduri, A. (2014). Unsafe Driving Detection System using Smartphone as Sensor Platform, *International Journal of Enhanced Research in Management & Computer Applications*, Vol. 3, pp. 65–70.

Galarza, E., Egas, F., Silva1, F., Velasco, P., Galarza, E. (2018) Real Time Driver Drowsiness Detection Based on Driver's Face Image Behavior Using a System of Human Computer Interaction Implemented in a Smartphone, *Proceedings of the International Conference on Information Technology & Systems (ICITS 2018)*, pp. 563–572.

Hasenjager, M., & Wersing, H. (2018). Personalization in advanced driver assistance systems and autonomous vehicles: A review, *IEEE Conf. Intell. Transp. Syst. Proceedings, ITSC*, Mar. 2018, 1–7.

Johnson, D. & Trivedi, M. (2011). Driving style recognition using a smartphone as a sensor platform, *Intelligent Transportation Systems (ITSC) 14th International IEEE Conference*, pp. 1609–1615.

Kalra, N., Bansal, D. (2014). Analyzing Driver Behavior using Smartphone Sensors: A Survey, *International Journal of Electronic and Electrical Engineering*, Vol. 7, pp. 697–702.

Kashevnik, A., Pilla, F., Russo, G., Timoney, D., Sweeney, S., Shorten, R., Ordonez-Hurtado, R. (2018). Context-Based Cyclist Intelligent Support: an Approach to e-Bike Control Based on Smartphone Sensors, *Internet of Things,*

Smart Spaces, and Next Generation Networks and Systems, Lecture Notes in Computer Science, Springer, Heidelberg. Vol. 11118, pp. 16–22.

Kashevnik, A., Karelskaya, K., & Repp M. (2019). Dangerous Situations Determination by Smartphone in Vehicle Cabin: Classification and Algorithms. Proceedings of the 24th Conference of Open Innovations Association FRUCT, *24rd Conference of Open Innovations Association FRUCT (FRUCT24)*, Moscow, Russia, 8-12 April 2019, IEEE. 2019. pp. 130–139.

Khattak, Z., Fontaine, M., Boateng, R. (2018). Evaluating the impact of adaptive signal control technology on driver stress and behavior using real-world experimental data, *Transportation Research Part F: Traffic Psychology and Behavior*, Vol. 58, pp. 133–144.

Kumar, A. & Patra, R. (2018). Driver Drowsiness Monitoring System using Visual Behaviour and Machine Learning, *IEEE Symposium on Computer Applications & Industrial Electronics (ISCAIE)*, pp. 339–344.

Kumar, A., & Patra, R. (2018). Driver Drowsiness Monitoring System using Visual Behaviour and Machine Learning. *2018 IEEE Symposium on Computer Applications & Industrial Electronics (ISCAIE).*

Lefevre, S., Carvalho, A., Gao, Y., Tseng, H.E., & Borrelli, F. (2015) Driver models for personalised driving assistance. *Vehicle System Dynamics*, vol. 53, no. 12, 1705–1720.

Ponomarev, A., & Chernysheva, A. (2019) Adaptation and Personalization in Driver Assistance Systems. *Proc. of the 24th FRUCT Conference*, April 8–12, 2019.

Sandeep, K., Ravikumar, P., Ranjith, S. (2017). Novel Drunken Driving Detection and Prevention Models Using Internet of Things, *International Conference on Recent Trends in Electrical, Electronics and Computing Technologies (ICRTEECT)*, pp. 145–149.

Van Ly, M., Martin, S., Trivedi, M. (2013). Driver classification and driving style recognition using inertial sensors, *IEEE Intelligent Vehicles Symposium (IV)*, pp.1040–1045.

Verkruysse, W., Svaasand, L., Nelson, J. (2008). Remote plethysmographic imaging using ambient light, *Opt Express*, Vol. 16(22), pp. 21434–21445.

Verwer, S., de Weerdt, M., Witteveen, C. (2011). Learning Driving Behavior by Timed Syntactic Pattern Recognition, *Proceedings of the 22 International Joint Conference on Artificial Intelligence*, pp. 1529–1534.

Wang, J., Yu, C., Li, S.E., & Wang, L. (2016). A forward collision warning algorithm with adaptation to driver behaviors. *IEEE Transactions on Intelligent Transportation Systems*, vol. 17, no. 4, 1157–1167.

Weidner, W., Transchel, F., & Weidner, R. (2017). Telematic driving profile classification in car insurance pricing. *Annals of Actuarial Science*, vol. 11, iss. 2.

Wu, M., Zhang, S., Dong, Y. (2016). A Novel Model-based Driving Behavior Recognition System Using Motion Sensors, Sensors, Vol. 16(10), pp. 1–23.

CHAPTER 4

INNOVATIONS IN HEALTHCARE AND EHEALTH

EDITOR: SERENA BAIOCCO

Abstract. Visual analysis and interpretation represent the gold standard in clinical practice. Therefore, the success of any clinical evaluation depends on the skills of the examiner and involves subjective variability. In the age of personalized medicine, more objective methods are needed to improve diagnosis, disease characterization, prognosis, and monitoring of therapy. Image analysis and processing constitute essential tools that can provide non-invasive or minimally invasive alternatives to improve, quantify, and maximize the *in vivo* information obtainable in clinical practice. This information plays a fundamental role in clinical decision making.

CHAPTER 4.1

STUDYING THE INFLUENCE OF NASAL CAVITY SHAPE ON AIRFLOW BEHAVIOUR DURING BREATHING

GENNADIJ LUKYANOV[1], ANNA RASSADINA[1],
ROMAN NERONOV[3]
1-ITMO University
gen-lukjanow@yandex.ru, a.a.rassadina@gmail.com
3- JSC "Modern Medical Technologies", Saint-Petersburg, Russian Federation
nrvspb@mail.ru

Abstract. The impact of the shape of the nasal cavity on respiratory physiology remains a subject of study. A method for measuring respiratory parameters directly in the nasal cavity, as well as its implementation through a sensor system, is presented here. The signal from a pressure sensor at the inlet of the nostril is used as the basis for digital simulation of air movement through the boreal paths. Pressure fluctuations in an opaque solid-state 3D model of a patient's nose have also been studied. Visualization of the airflow in nose cavities has been realized through experiments with transparent models, as well as through digital simulation. Visualization experiments on solid and digital models can show differences in tidal exchange depending on the shape of the nasal cavity. Experiments have shown that in a wide nose, when inhaling and exhaling, the air tends to move through the upper areas of the nasal cavity. During inhalation and exhalation in a narrow nose, the air moves along the bottom and middle parts of the nasal cavity. In the model of the middle nose, during inhalation the main part of the airflow is concentrated in the lower lobes, while during exhalation the air is evenly distributed between the middle and lower regions.

I. Introduction

The nasal cavity has a branched shape and irregular cross-section. Turbinate bones divide the cavity into four nasal passages (Fig. 1). The complex internal structure of the nasal cavity makes the study of breathing difficult.

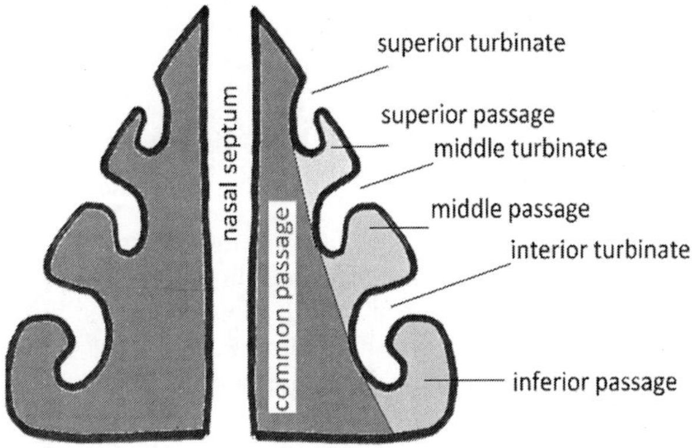

Fig. 1. Coronal slice of nasal cavity with the division into nasal passages

Respiration studies follow three avenues of research. Experimental studies on full-scale models of the nose, using digital simulations, and on living material, including humans and animals, have been undertaken. Hybrid research has been done using the nasal cavities of corpses.

The first full-scale models of airflow through the nasal cavity were produced in the 1920s and 1930s (Mink P., Hofbauer, Takahashi K., Businger O., Scheideler J.) [1-3]. As a rule, these analyses only involved visual observations of stained liquid or smoke moving through a solid model; the models could be increased in size for more convenient observation. As such, the authors [2-3] used rectangular-shaped boxes, inside of which it was possible to create motion interference defining the air's flow angle and its movement.

Numerical models give the best visualization. Using such models, one can see the distribution of velocity, temperature, and pressure fields at any section of the nasal cavity. The advent of numerical models is a consequence of the development of computer and medical technology. These developments include the digitization of topographic imaging data, the tremendous growth in the 1990s of computing power, and the emergence of computing software for visualizing aerodynamic and hydrodynamic processes. The first 3D-model of a half-truncated nasal cavity in the nasopharynx area was created in 1995 (Keyhani et al.) [4]. In recent decades, we have seen an increase of work on this theme [5-7]. Depending on the variability or constancy of the modelled flow parameters and time, it is possible to identify two large groups with stationary (independent of time) or unsteady flow (considering changes of flow in time) simulation. Stationary models use the mass airflow rate [5] or pressure drop [6] as a boundary condition. Unsteady models also use boundary conditions, like stationary models, with the conditions of fixed-rate flow at the entrance to the nasal cavity [7].

When performing research in humans, the main difficulty is the impossibility of placing diagnostic sensors directly in the nasal cavity. Having to take measurements in a tube leading to the nasal cavity and sometimes to the mouth is the main drawback of current diagnostic methods. In today's medical diagnostic equipment, the measurement of the physiological parameter of respiration (nasal resistance) is determined by the ratio of pressure drop to airflow in the tube supplied to the nasal cavity, undertaken with an anterior active rhinomanometer. These provide indirect measurements that are used in the method of acoustic rhinometry. However, this problem can be solved by applying miniature high-speed sensors. The sensitive elements of those sensors can be placed in the nasal vestibule. As such, we have developed a number of prototypes, the construction and operation of which are given in [8-10].

The movement of airflow within the nasal cavity is an open topic and research in this area will expand our knowledge of respiratory physiology. This is useful for surgical intervention, where it is very important to understand airflow behaviour. Since modern medicine seeks to minimize harm, it is necessary to understand how air should move in the nasal cavity during the process of breathing.

The results of the proposed research are to be used as the basis for the implementation of a *virtual operation* program. It is assumed that mathematical modelling of upcoming surgery will be performed on a 3D model of the patient's nasal cavity that has been created using computed tomography (this is the *virtual operation*). This operation will help minimize the impact of surgery by increasing understanding of the features of a particular nasal cavity's anatomy and its influence on the movement of air in the cavity during breathing (i.e., understanding its norms and pathologies).

Pathways of air movement in the nasal cavity during breathing are extremely individual. There are three evolutionarily formed shapes of the nasal cavity: platyform, mesoform, and leptoform.

Previous scientific research has shown that differences in airflow movement depend on the shape of the external nose [11-15]. However, the shape of the external nose does not always correspond to the shape of the nasal cavity. For this reason, we carried out a study of air movement in leptoform, mesoform, and platyform cavities during breathing. The shape of the nasal cavity was determined by the method found in [16]. The effect nose shape on air movement is a one crucial task.

The aim of this work is to estimate the airflow moving through solid 3D models of leptoform, mesoform, and platyform nasal cavities during a simulation of breathing. The results of the study and the numerical simulation of airflow will be compared to existing studies on human subjects. A new method of measuring parameters of respiration in the nasal cavity and prototypes of sensor devises will also be presented.

This information can be used to refine existing models of these processes, which, in turn, will allow us to better understand the evolution of the processes of inhaling and exhaling and present a more accurate assessment of the human condition.

Numerical and solid 3D nasal cavity models were constructed using computed tomography (CT) data. Solid 3D models were created using 3D-printer technology at a scale of 1:1 and using the same CT data. Digital models were used for simulation in the ANSYS package. Air was considered to be an incompressible quasi-viscous liquid for modelling purposes. The motion of airflow was described using partial differential equations (Navier-Stokes) [17].

II. Materials and Methods

A. Method of measuring respiration parameters

Two generally accepted methods can be used: anterior active rhinomanometry and acoustic rhinometry.

As noted earlier, the main weakness of anterior active rhinomanometry is the necessity of reading data from a tube. This sees the loss of important diagnostic data and the distortion of natural breathing. The method of acoustic rhinometry is indirect. This method allows us to estimate the cross-sectional area of the airway channel. Diagnostic parameters measured directly in the nasal cavity give significant data.

Previously [8, 9, 17], it was shown that any parameter reflecting air oscillations in the nasal cavity during breathing could be chosen as a diagnostic parameter. These could include airflow velocity, pressure, temperature, or even, for example, the change in carbon dioxide concentration.

The following provisions were proposed for the development of the sensor system for measuring respiration parameters: (1) sensor elements should be located in the nasal cavity, without using tubes that distort the readings; (2) sensor elements should be capable of rapid operation.

We developed several sensor devices [8-10, 17] with sensor elements located directly in the nasal cavity without using tubes that distort the readings. These can be thermistors, thermocouples, metres, pressure sensors, optocouplers, and other sensors.

As such, the sensor device shown in Fig. 2a allows us to measure the temperature and the pressure of the nasal cavities without additional tubes.

The sensor elements were placed on a miniature clip. The clip could be attached to the nasal septum (in the nasal vestibule). A type K (Chromel-Alumel) thermocouple was used for temperature measurement. The thermocouple was welded together from wires with a diameter of 0.08 mm. This gives a response time of some centi-seconds. Two sensors were applied for pressure detection. The pressure range of the sensor was ~4 kPa. Pressure control was realized by tubes with an inner diameter of 0.5 mm. The main advantage of the sensor is the preservation of the natural breathing process.

Amplifier and normalization blocks were used to increase gain and normalize the signals from the thermocouples and the pressure sensors. Software to write data onto a hard disk drive was also developed for the sensor.

Fig. 2b presents a miniature sensor device for measuring the fluctuation of CO_2 concentrations during breathing. It is known that the concentration of CO_2 changes during respiration: the proportion of CO_2 at inhalation is 0.03 %, while at exhalation it is 4.00 %. This parameter is important in breath diagnostics and is very informative. The basic absorption band of CO_2 corresponds to the spectral region between 4,200 and 4,320 nm. Weaker absorption bands are located at 2,700 nm and 2,000 nm. In our study, we used an optocouple with dominant wavelengths at 3.4, 3.8, and 4.1 μm at a temperature of 300 K.

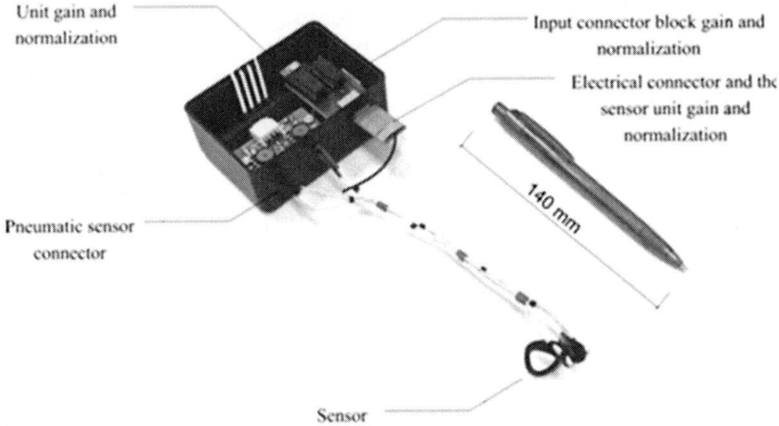

a

b

Fig. 2. Miniature sensor system with measurement elements: (a) thermistors; (b) thermocouple; (c) optocouple

Studies performed with prototypes of the sensor devices showed the significance of weak fluctuations, which have been observed to display oscillation in the measured parameter during the processes of inhalation and exhalation.

A new diagnostic method was developed based on nonlinear analysis of airflow oscillation in the nasal cavity. Firstly, the power of the spectral density of the measured signal needed to be chosen. This can be calculated by use of the discrete Fourier transform:

$$y(k) = \sum_{m=1}^{n} x(m) e^{-j2\pi\frac{(m-1)(k-1)}{n}} \quad (1)$$

where j is the designation for imaginary unit; n is the number of elements of a given vector; x, y provide the output vector; and m is the measured value of the signal.

It is known that all biological processes are dissipative, i.e. a phase trajectory can be constructed for them [18]. A matrix composed of three columns, represented by the relevant vectors, was used to build the phase trajectory in a three-dimensional coordinate system. The first column vector is the measurement of the signal over time: $x(t)$. The second column vector is the measurement of this signal with a single shift in time $x(t+\tau)$, where τ is a one point time-shift. The third column vector is the same measurement result, but with a time shift of 2 points $x(t+2\tau)$. The phase trajectory was constructed with coordinates where values without a time shift are plotted vertically and values with a time shift, τ and 2τ, are plotted along the two horizontals. The points on this graph will tend to display a certain range in space, forming an attractor.

In our study, the graphs of the spectral power density and phase trajectory of the measured signal were plotted with the mathematical modelling package of Matlab© (MathWorks, Natick, MA, USA).

B. Generation of prototype models

The prototype models were created using a geometric model of human internal nasal channels with leptoform, mesoform, and platyform cavity type noses. These models provide the basis for numerical modelling and 3D printing of solid models. They were reconstructed using patient CT scans at the *Modern Medical Technologies Centre* (St. Petersburg).

The data for each type of nose comprised a set of images of cross-sections of the human cranium with a step interval of 0.625 mm. The parts corresponding to the air inside the cranium were isolated from the image slices. Furthermore, the specified segmentation parameters of the entire set of slices were summarized and superimposed on each other. This allowed the construction of a 3D image of the internal nasal canals.

Fig. 3 shows an image of a 3D geometric model for the mesocavity type of nose. This geometric model was segmented with *entrance*, *exit*, and *wall* boundaries (Fig. 3b). The boundaries were defined with the help of the Altair Hypermesh software package. Each geometric model was then imported along with the segmentation parameters using the Ansys software package to an STL file (Fig. 3b). In Ansys Icem CFD, an irregular surface and volume finite element mesh were sequentially constructed with the following parameters:

- The maximum size of the final element edge was 0.4 mm. This choice was driven by: (1) the minimum gradation value of the tomograph; (2) the capabilities of the computing resources available.
- The values of $4.5 \cdot 10^6$ for the triangular surface elements and $3.9 \cdot 10^7$ for the bulk tetrahedral elements.

Numerical modelling was performed using Ansys Fluent software.

More detailed information about airflow motion in the nasal cavity was achieved by using two models at once. In order to perform experiments, two types of solid 3D model were created.

- Opaque models that fully reproduce the internal structure of the nasal cavity.
- Transparent models of the right half of the nasal cavity without the maxillary sinus.

The opaque models were used to experimentally measure airflow parameters. Here, pressure fluctuations were taken as a parameter of airflow.

The transparent models were used experimentally to visualize airflow in the nasal cavity.

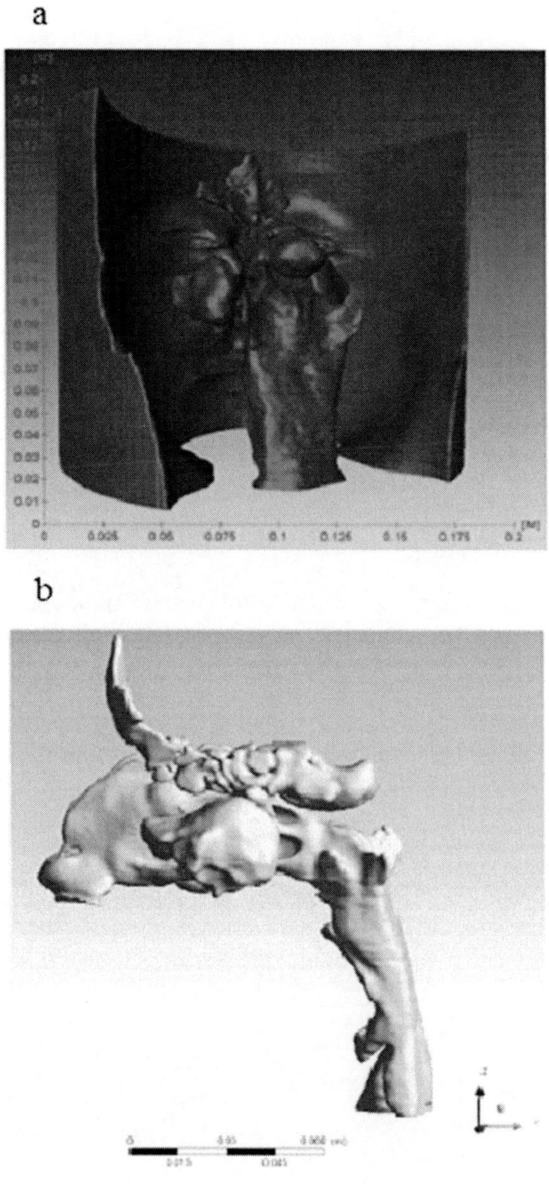

Fig. 3. Geometric models: (a) base model; (b) edited model

C. *The numerical model*

It is known that during the respiratory cycle airflow is unsteady in the nasal cavity. A conversion from laminar flow to turbulent vortices and back is seen. This is all determined by the structure of the nasal cavity, which has areas with flow separation and recirculation zones. In this case, a non-stationary calculation should be performed to obtain data on turbulent pulsations.

There are three main groups of mathematical model, the use of which is possible in the implementation of numeral experiments [19]:

- RANS model (Reynolds-averaged Navier Stokes): mathematical models based on the solution of Navier-Stokes time-averaged equations.
- LES model (large eddy simulation) and its derivatives (DES, DDES, IDDES etc.): suggesting a direct numerical calculation of predetermined value turbulent flow parameters.
- DNS model (direct numerical simulation), based on the direct numerical solution of motion equations for all spatiotemporal characteristics with resolution of all turbulent structure scales within the flow.

A fuller analysis can be executed more completely through direct numerical simulation. However, currently the solution to this problem is severely restricted by the limits of modern computing power: in spite of the Reynolds numbers (Re), relating to the boundary values between laminar and turbulent regimes in the nasal cavities, mesh generation for the DNS method with the number of cells at the proportion Re^3 makes the values Re at 2×10^3. These are unacceptably high and the total number of mesh elements in this connection amount to the order of 10^{10}. Currently, the best solution to this problem is found in DES, DDES, and IDDES frameworks, and other derivatives of the LES model. In this case, the number of calculated elements in the mesh will be of the order of 3.5^7, which is usable in calculation.

From among the workable methods for calculating turbulent flows, the detached eddy simulation Spalart-Allmaras model (DES-SA) was chosen. Non-stationary calculation of the thermal and hydrodynamic parameters of airflow—velocity, temperature, and pressure—was made for the respiratory cycle.

Navier-Stokes equations were used to describe fluid flow. The equation system includes a continuity equation and equations for flow velocity.

The continuity equation is

$$\frac{\partial \rho}{\partial \tau}+\frac{\partial \rho u_j}{\partial x_j}=0 \tag{2}$$

The motion equation is

$$\frac{\partial \rho u_j}{\partial \tau}+\frac{\partial \rho u_j u_i}{\partial x_j}=\frac{\partial P}{\partial x_i}+\frac{\partial \tau_{ij}}{\partial x_j} \tag{3}$$

where, u_i, u_j are the flow velocity components; ρ is the density; P is the pressure; and τ_{ij} is the deviatoric stress tensor.

Equations (2) and (3) contain the variables u_i, u_j representing the instantaneous velocity of flow. The change of these parameters in space and time in a turbulent flow regime is so complex that they cannot be calculated exactly. For this reason, all currently used mathematical models of turbulence imply a number of simplifications of the system of equations in (2) and (3). The most commonly used is Reynolds averaging, in which the instantaneous values of velocities and pressures are replaced by the sum of their averaged and pulsating values. There are various modifications of such averaged equations that are used as the basis for mathematical models of turbulence: the k–ε and k–ω models [20] and the hybrid Menter model [21], among others. The principal disadvantage of these models is the need for initial averaging of pulsating quantities (velocity, pressure) over the entire frequency spectrum of oscillations, which excludes the possibility of modelling small-scale turbulent vortices by means of which the kinetic energy of the flow dissipates. This disadvantage can be eliminated using the LES and DES methods. The basis of these methods is direct numerical calculation of the desired parameters of turbulent flow within the scaled framework specified by a special filter. The resolution of scales smaller than the filter size occurs through approximate modelling. As a rule, a finite element grid is used as a filter. Thus, the part of the frequency spectrum of turbulent oscillations of the unknown quantities is calculated exactly using equations (2) and (3).

The smallest vortices, whose exact calculation are not technically feasible, are modelled through averaged values of the velocities and pressures of the flow.

The most significant advantage of the DES algorithm over the RANS method is the possibility of gathering information on high frequency eddies inside the flow, which is gained by replacing time filtration (RANS models) with spatial filtration (DES methods). In this study, air movement modelling was carried out using the mathematical DES model. Scales smaller than the grid size were modelled using the Menter hybrid model [21].

The mathematical formulation of velocity field vector filtration $\mathbf{u}(\mathbf{x},t)$ is

$$\overline{\mathbf{u}}(\mathbf{x},t) = \int \mathbf{G}\left(\mathbf{r},\mathbf{x}\right)\mathbf{u}\left(\mathbf{x}-\mathbf{r},t\right)d\mathbf{r} \tag{4}$$

where $\mathbf{G}$ ($\mathbf{r}$, $\mathbf{x}$) is the filtration function; $\mathbf{x}$ presents the vector of coordinates in Cartesian space; $\mathbf{r}$ represents coordinate points for vector outlets from which integration is performed; and t is non-dimensional time.

The width of the filter determining the function $\mathbf{G}$ ($\mathbf{r}$, $\mathbf{x}$) can be determined explicitly, or according to the size of the finite element mesh. Equations (2) and (3) filtered can be simplified to the following form according to equation (4)

$$\begin{aligned}\frac{\partial\rho\overline{u}_i}{\partial\tau}+\frac{\partial\rho\overline{u}_i\overline{u}_j}{\partial x_j}&=-\frac{\partial}{\partial x_i}\left(\overline{p}+\frac{2}{3}\mu\frac{\partial\overline{u}_k}{\partial x_k}\right)+\\&+\frac{\partial}{\partial x_i}\left(\mu\left(\frac{\partial\overline{u}_i}{\partial x_j}+\frac{\partial\overline{u}_j}{\partial x_i}\right)\right)-\rho\tau_{ij}^{r}\end{aligned} \tag{5}$$

where τ^{r}_{ij} is the stress deviator tensor. This parameter must be determined in order to solve equation (5).

Simulation of the stress deviator tensor is performed using eddy-viscosity sub-grid-scale models, such as the Smagorinsky model. This model rests on the hypothesis of proportionality between the stress deviator tensor and the strain velocity tensor of the filtered area; it is structurally similar to the Boussinesq hypothesis.

In accordance with the DES method, the linear scale l_{RANS} is replaced with the following ratio:

$$l_{DES} = \min\left\{l_{RANS}, C_{DES}\Delta\right\} \tag{6}$$

where l_{DES} is the DES linear scale; C_{DES} is constant; Δ is a filter size and is determined as:

$$\Delta = \max\left\{\Delta_x, \Delta_y, \Delta_z\right\} \tag{7}$$

where Δ_x, Δ_y, Δ_z is the finite element size.

The result is the congruence $l_{DES} = l_{RANS}$ and the DES model functions as a RANS model, if the size of the filter (computational mesh) is too crude and $C_{DES}\,\Delta > l_{RANS}$.

Otherwise, the DES model is transformed into an LES model and complies with the solution of the filter equations (5). As such, the DES method allows us to reach a significantly transient solution to the problem.

The following boundary conditions were assumed:

The pressure in the vestibule of the nose (*entrance*)

$$P = 0\ Pa, T = 293\ K \tag{8}$$

The spatial components of airflow velocity on a *wall*

$$u_1 = u_2 = u_3 = 0 \text{ m/s}, T = 310\ K \tag{9}$$

The pressure in the nasopharynx (*exit*) at inhalation

$$P = 50 \cdot \cos\left(\pi\left(0.87\tau + 0.5\right)\right) + 50, Pa, T = 310\ K \tag{10}$$

the pressure in the nasopharynx (*exit*) at exhalation

$$P = 50 \cdot \sin\left(\pi\left(1.18\tau + 0.5\right)\right) - 50, Pa, T = 310\,K \qquad (11)$$

A non-stationary calculation was carried out for a full inhalation-exhalation cycle of 4 s duration, in steps of 0.1 s.

The advantage of this model is its visualization of trajectories of small volumes of air in the general structure of the airflow. This allows us to identify individual vortex components of turbulent flow and thus obtain a more detailed picture of the physiology of respiration. The question remains as to how accurate the given boundary conditions are? We took these boundary conditions into account in the research we performed using solid 3D models and diagnostic human studies.

D. Experimental modelling

The tasks of experimental modelling included:

- obtaining pressure measurements in the path of the airflow;
- visualization of the flow along its movement in the nasal cavity.

In the first instance the experiment was performed on an opaque model. In the second instance, we used transparent models. Let us present a more detailed description of experimental installations here.

1) *Opaque models*. These models were developed by 3D printing at a scale of 1:1. The models completely replicated the nasal cavity, including all the nasal sinuses that open into the nasal cavity.

Air motion imitating breathing was performed using a tube fixed to the nasopharynx area of the model. The experimenter carried out successive inhalations and exhalations through the tube during the simulation of respiration. As such, the simulation can be considered to approximate human breathing.

The setup for performing an experiment on an opaque model includes a sensor system and a data automation subsystem.

Pressure sensing was developed with piezoresistive pressure sensors. The piezoresistive pressure sensors were connected through thin shunt tubes to multiple locations on the solid model for fixing pressure drops in the airflow in time. The installation of shunt tubes in the correct locations

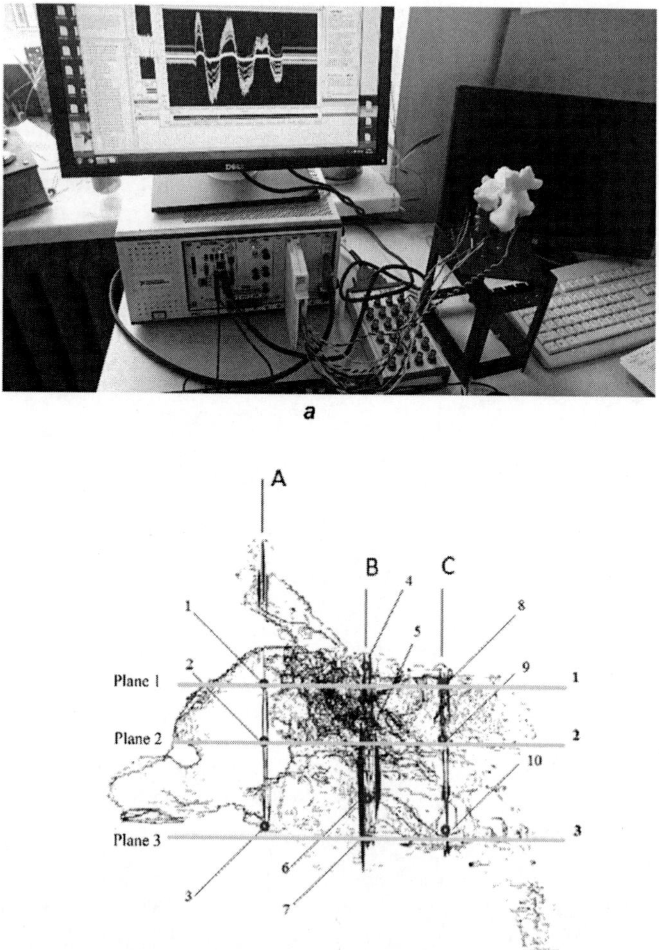

Fig. 4. Sensor system for breath parameter measurement: (a) general view of the measurement setup with results; (b) locations of thin shunt tubes for measuring pressures in the artificial solid model of the human nose cavity

of the solid-state model was performed at the *Modern Medical Technologies* clinic with the help of an endoscope. The measuring system and locations of the shunt tubes in relation to the pressure sensors are shown in Fig. 4.

Nine sensors were located in the common nasal passage: in the upper, middle and lower planes and A, B, C sections of the nose cavity (sensors 1-5, 7-10). One (sensor 6) was located in the middle nasal passage section B. This placement of sensors allowed coverage of all major areas of air movement in the nasal cavity.

The signals received from the pressure sensors were supplied to the data automation subsystem, which controlled and synchronized the sensors. Settings and data recording was performed using software provided by the manufacturer. Data processing was performed with the Matlab mathematical modelling package. Data processing was performed using in-house software to perform the conversion of the voltage output values of the sensors in pascal (Pa), filtered data, and deleted steady components at the pressure transducer readings, displayed by the sensors as a default, depending on the time domain.

The pressures from the right and left halves of the nasal cavity of the solid 3D model were measured during the experiment.

2) *Transparent models*. These models were created using 3D-printer technology at a scale of 1:1, using the same CT data, in polished plastic. These models only copy the shape of the nasal cavity. Maxillary sinuses are not represented in the model.

The transparent models are presented in Fig. 5.

For the experiment, the breathing was modelled as follows: an airflow with a temperature of 35 °C and a peak flow rate of 10 l/min was provided using a PVC tube inserted at the nasopharynx. During the modelling process of breathing, the experimenter breathed through a tube attached to the nasopharynx of the model.

The airflow distribution was estimated using the Testo 890 thermal imager in the temperature range of 20-30 °C. The experimental setup is shown in Fig. 6.

The method used to study airflow movements was tested before its application.

The infrared imaging method was tested using a sensor system. This system included four rapid response copper/constantan thermocouples and a data acquisition device (Fig. 7). A data acquisition system was used for storing and recording the measurements from the thermocouples.

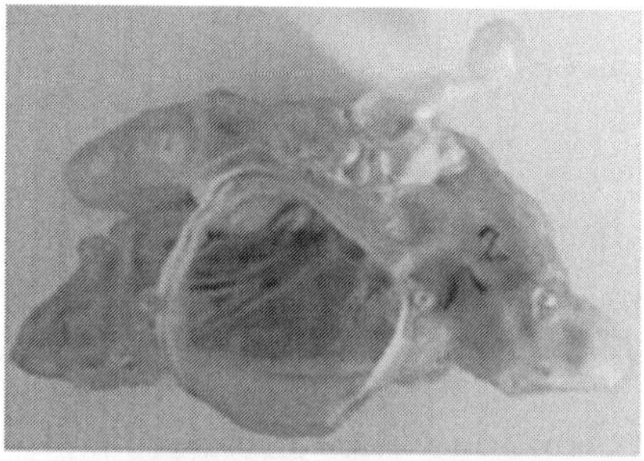

a

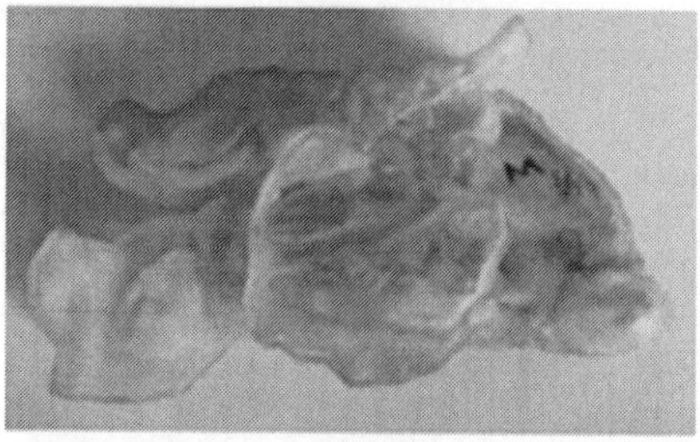

b

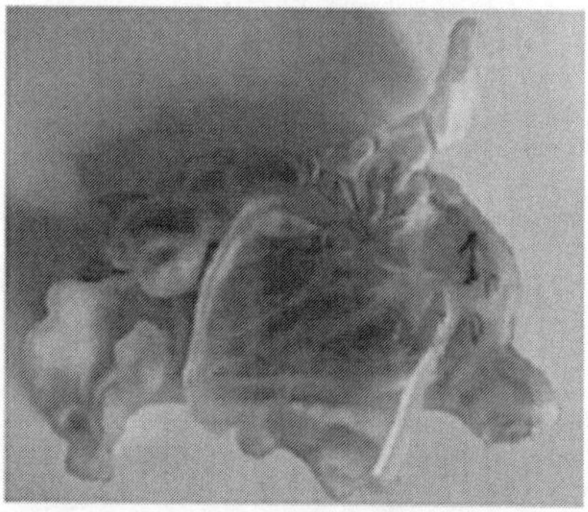

c

Fig. 5. Solid 3D models of the right cavity of a nose: a) leptocavity; b) mesocavity; c) platycavity

Thermocouples 1, 3, and 4 were installed through a fine diameter bore (0.5 mm) at the inner wall of one of the transparent models. Thermocouple 2 was located at the outer wall of the model. Note that, since the experiment aimed to identify the pattern relationships between the readings from the thermal imager and the readings at the thermocouples, the choice of model type (platyform, mesoform, or leptoform) was not fundamental here. Simultaneous temperature field measurements were performed using thermocouples and a thermal imager. The results were compared.

Data obtained from the thermal imager and four thermocouples were processed using the Matlab software package. Based on the results obtained, the following conclusions can be drawn:

1. The data of the thermal imager correspond to the thermocouple fluctuations at the given points (Fig. 7c).
2. The inhalation-exhalation period was 16 s.
3. The time delay between the results of the thermal imager and the thermocouples 1, 3, and 4 was 2.4 s. These thermocouples were located inside the model. This time delay is less than a quarter of a period and confirms that the wall heats up and cools quickly.
4. The time delay between the results of the thermal imager and thermocouple 2 was 0.1 s. The thermocouple was located on the outer surface of the wall. This thermocouple has a minimal time lag between the readings of the thermal imager and the thermocouple. The periods of wall heating and cooling are almost identical.
5. Comparing the inhalation-exhalation period and the time delay, shows that the wall of the model quickly heats and cools. As such, we can conclude that the readings of the imager in the simulation of respiration will reflect fluctuations in the temperature field near the nasal septum of the solid 3D nasal cavity model.

III. Results

E. Results of studies on living material

The main feature of our method is the placement of diagnostics sensors in the nose without the use of additional tubes, as is the case in the active anterior rhinomanometry method. Regardless of the choice of diagnostic sensor, the signal from the measuring elements is identical in shape. This can be seen in Fig. 8a.

The diagnostic method can be based on an analysis of the graph showing the spectral power density of the measured signal. A healthy person displays two characteristic peaks. One of these corresponds to the frequency of respiration, while the second one corresponds to high-frequency oscillations superimposed onto the respiratory process. These weak fluctuations are caused by the formation of a small scale air eddy in the nose, with a change in the direction of the air, upon shifting from inhaling to exhaling and vice versa. The breathing of a sick person is confused and arrhythmic. This is immediately noticeable as the number of peaks increases. The analysis of the phase trajectory of these signals is also very indicative. In our study, the graphs of the spectral power density and phase trajectory of the measured signal were plotted using the Matlab mathematical modelling package.

One can see that breathing is a dynamic process. External respiration is characterized by fluctuations in inhaled and exhaled airflow. These fluctuations have a non-linear dissipative character and in practice can be described by any variable of airflow. We investigated the temperature and the pressure fluctuations of airflow in the entrance of a nose in a healthy person and in a person with asthma. Our studies allowed us to detect small-scale turbulent eddies. We estimated eddies through an analysis of the power spectral density and by the reconstruction of the attractor in phase space.

Fig. 6. The experimental setup for thermographic research

Fig. 8b shows a graph of the spectral power density of the measured signal. Fig. 9 presents the phase trajectory of these signals. It is easy to see that the phase trajectory of a sick person is heavily tangled. The moment of inhalation and exhalation merges with one another. At the same time, the moment of inhalation and exhalation is clearly visible in the phase trajectory in the results of measuring a healthy person's respiration.

F. Analysis of experimental modelling results of an opaque model

The input signal obtained by the prototypes sensor devices was used as the basis for numerical simulation. Fig. 10a shows the results of experimental measurements of pressure drops in a solid model of a human nasal cavity. The pressure drop was measured in the vestibule of the nose (nostrils). Fig. 10b shows the result of numerical simulation for the corresponding computer model of the nasal channels (vestibule of the nose). The results of digital simulation are correlated with experimental results. Note that the numerical calculation of non-stationary boundary conditions was given for *exit* (the region of the nasopharynx). The similarity of the waveforms and amplitudes of the computer and solid models does not allow us to make a conclusion about the adequacy of the computer model to the nasal channels and the method of computing the hydrodynamic problem.

The data obtained during the experiment were recorded in a text file. Further processing of the data was done in the Matlab mathematical modelling package. From this we obtained 10 graphs of pressure dependence on time and 10 graphs of the power spectral density for the right part of the solid model of the nasal cavity. A similar result was obtained for the right half of the nasal model.

Analysis of the results obtained for the right half of the nasal model showed that the amplitude values of pressure increased with the advance of air into the nasal cavity. The maximum values were in the upper 2/3 segments of the nasal cavity model.

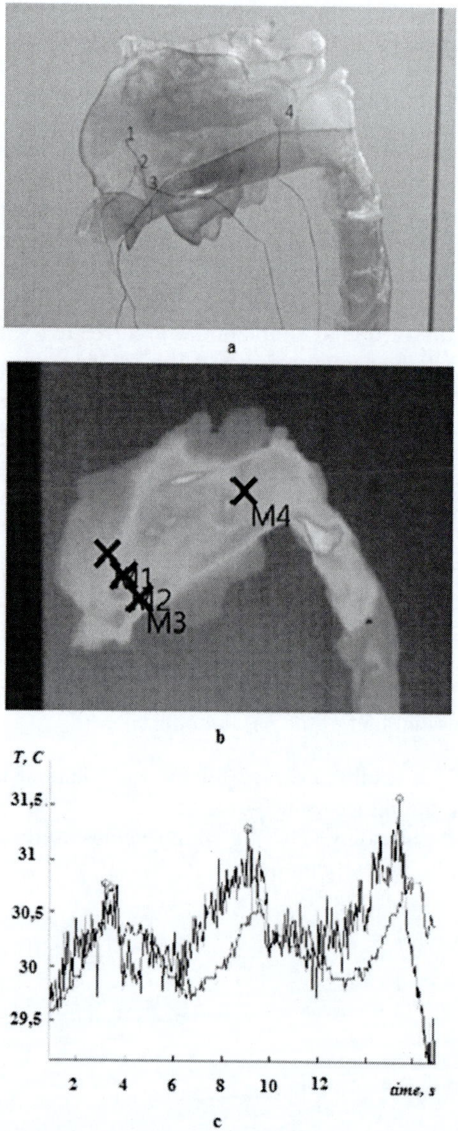

Fig. 7. (a) Location of thermocouples in the sensor system; (b) points for measurements with a thermal imager; (c) comparison of measurement results by thermocouple (black) and thermal imager (blue)

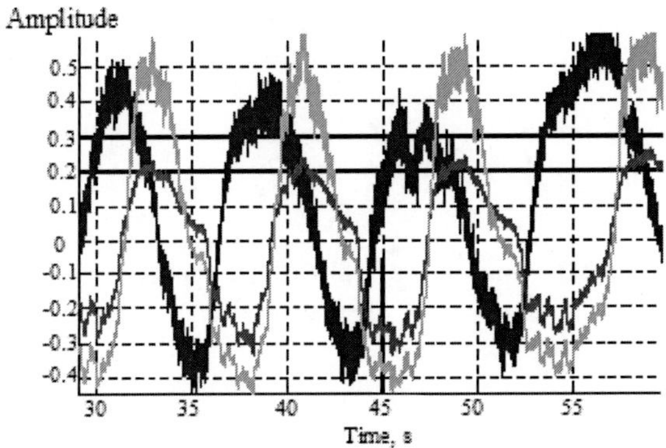

a

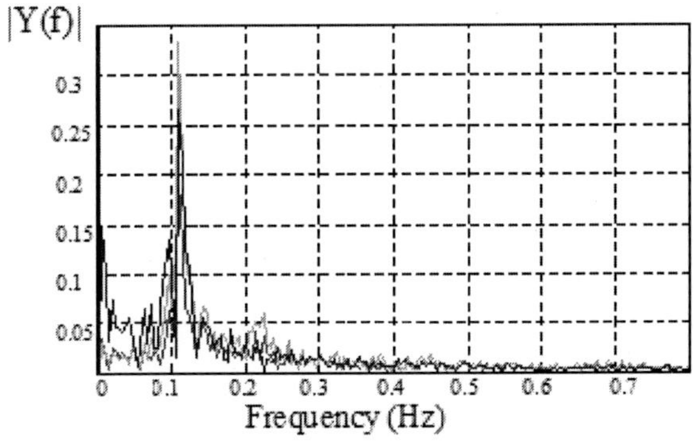

b

Fig. 8. (a) Results of the measurement of fluctuations of temperature and CO_2 concentration during respiration; (b) plots ot the power spectral density for the measurement results. Color key: blue: the concentration of CO_2; red: the results of pressure measurements in the left nasal cavity; green: the results of pressure measurements in the right nasal cavity

The results obtained for the right half of the model were somewhat different. At all points of the study, approximately identical amplitude pressure values were observed, with the exception of the upper third where the amplitude pressure values were lower. As such, airflow displays individuality even in the consideration of a single nasal cavity.

However, the analysis of graphs of the spectral density of the nose showed that both the left and right halves of the nasal cavity see two characteristic peaks at frequencies of 0.380 Hz and 1.1 Hz. The first frequency corresponds to the frequency of breathing in, and the second is the frequency of total cycle aspiration. This peak also coincides with the heartbeat.

The frequencies at which the peaks of the spectral power density were observed in the studies on the opaque model differ from the analogous frequencies observed in studies of living material. In this last case, the peak was about 1.12 Hz. In actuality, the inhalation-exhalation cycle was longer than in the experiments involving humans. In studies using an opaque model, the experimenter breathed through the tube supplied to the model. Involuntarily, the experimenter performed deeper breaths and exhalations, which led to a decrease in the frequency of breathing.

G. Comparison of measurement results in the numerical model and transparent models

The results of numerical simulation and thermography are shown in Figs. 11, 12, and 13.

In Fig. 11, with inhalation and exhalation one can see that the air predominantly flows through the lower regions of the model of the leptoform cavity.

Fig. 12 shows the results of measurements in the mesoform cavity. During inhalation, the main part of the airflow was concentrated in the lower lobes of the model, while in exhalation the air was evenly distributed between the middle and lower regions.

In the platyform cavity, during inhalation a movement of air to the upper sections is seen; while at exhalation the air is evenly distributed throughout the area of the nasal passage (Fig. 13).

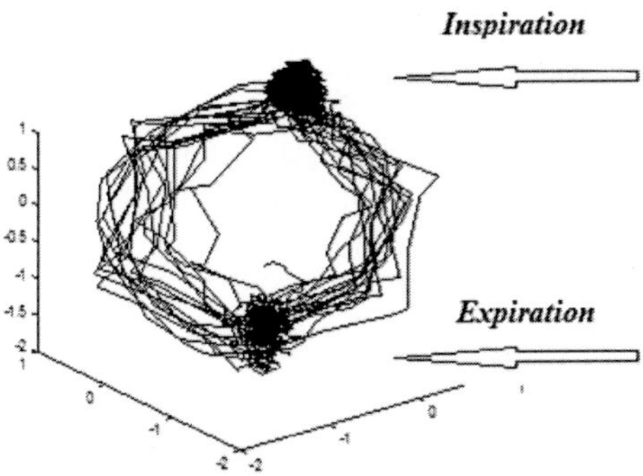

a

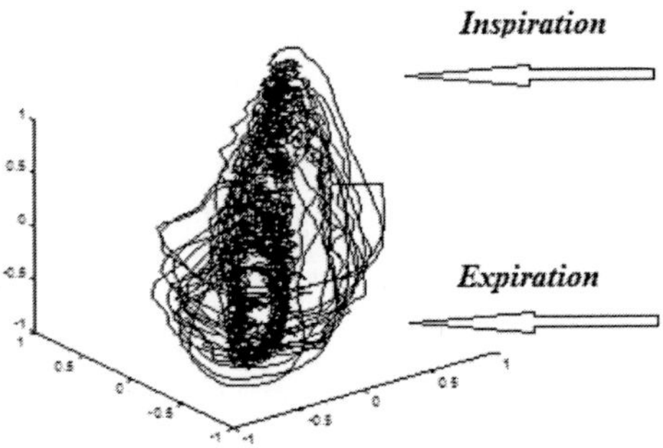

b

Fig. 9. The reconstructed attractor in the phase space for temperature fluctuations in the nostril: (a) of a healthy person; (b) of a patient with asthma

IV. Conclusion

The study of liquid or gas flow in irregularly shaped channels is a non-trivial task. Such channels, the example of the nasal cavity here, are characterized by the presence of a large number of areas with flow separation and the presence of recirculation zones. Studies are hampered by the small dimensions of such channels. At the same time, understanding the physiology of breathing is at the cutting edge of rhinosurgery. An integrated approach was taken during the performance of this study. A new method with a novel sensor device was designed to study the physiology of human respiration. This method is based on consideration of respiration physiology as a dynamic process. Two types of simulation, numerical and experimental, were performed. The input signal obtained by the sensor prototypes was taken as the basis for numerical simulation. A feature of numerical simulation is the implementation of non-stationary calculations using a DES model that, on the one hand, allows the observation of small eddies in the airflow, and on the other, gives an acceptable computational grid size for calculation (about 3.57 elements). Note that among other models of turbulence, RANS models do not allow the observation of small eddies in the airflow, while DNS models have a computational grid with a very large number of elements (about 10^{10}), requiring the calculations to be done on super-powerful computers. Since numerical simulation depends heavily on the choice calculations for the model, and on its initialization, the results of numerical simulation must be verified by experimental studies. Experimental studies using opaque models obtained data on pressure fluctuations at points within the solid model of the nasal cavity. On the basis of these data, the requirements for specifying the initial conditions when performing numerical simulations were formulated. Furthermore, a series of experiments was carried out with a thermal imager and transparent models of the right half of the nasal cavity. The visual images of the temperature field near the septum of the nasal cavity model were obtained. These results were compared with the numerical simulation data.

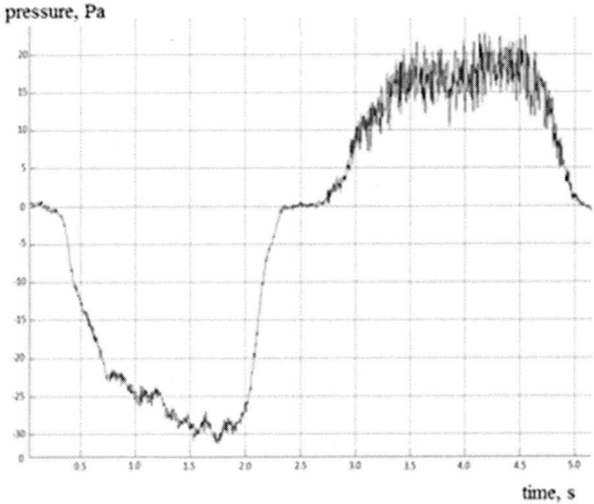

a

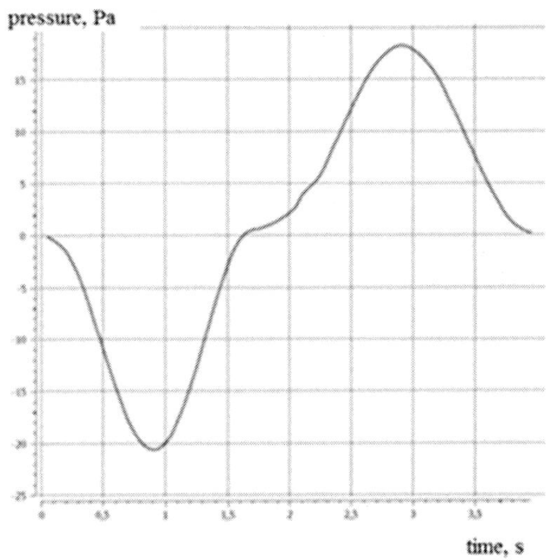

b

Fig. 10. Results of the measurement and modelling of pressure drops in the vestibule of the nose: (a) solid model; (b) numeral computation

Comparison of the results gives the following conclusions:

1) In a wide nose (platyform cavity), it should be noted that when inhaling and exhaling the air tends to flow through the upper areas of the nasal cavity.

2) In a narrow nasal cavity (leptoform cavity), during inhaling and exhaling the air moves along the bottom and middle parts of the nasal cavity.

3) In a medium-sized nose (mesoform cavity) during inhalation, the main part of the airflow is concentrated in the lower lobes of the model, while during exhalation the air is evenly distributed between the middle and lower regions.

Acknowledgments

We would like to thank our programmer Aleksey Voronin for his work and collaboration on numerical simulation. Our work would not have taken place without the participation of this wonderful specialist.

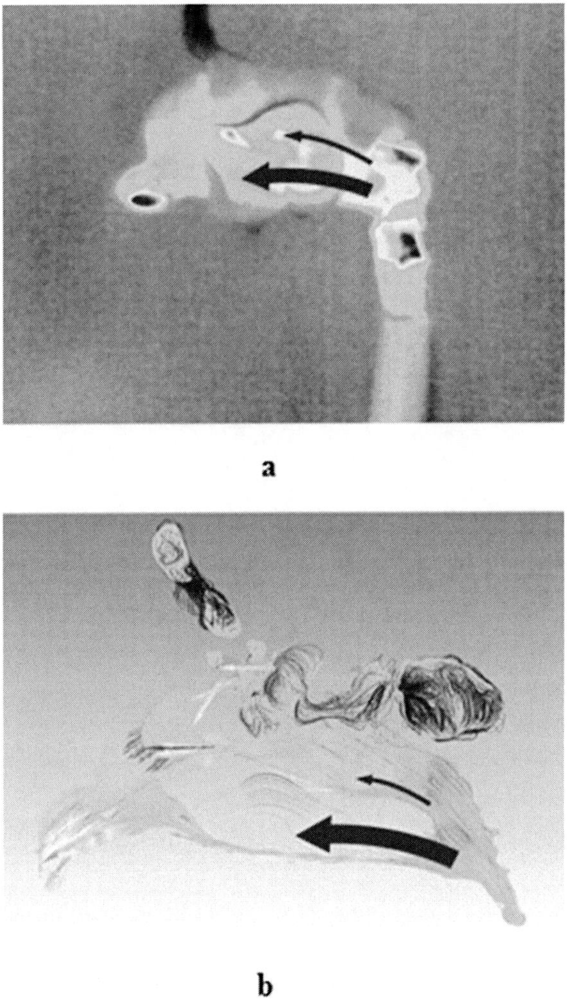

Fig. 11. Results of (a) thermography and (b) numerical modelling of a leptoform nasal cavity

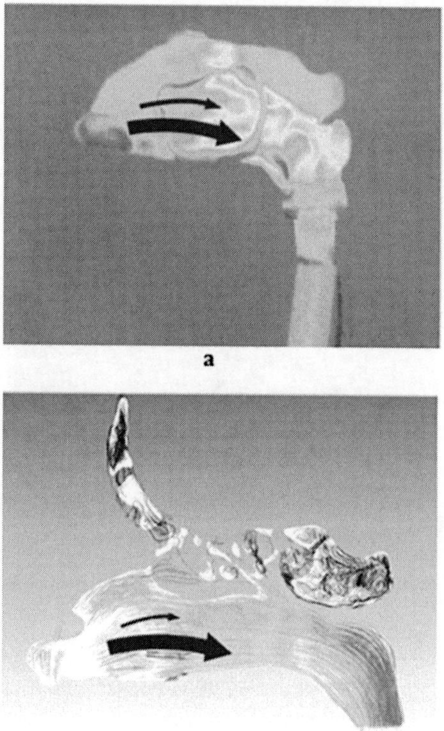

a

b

Fig. 12. Results of (a) thermography and (b) numerical modelling of a mesoform nasal cavity.

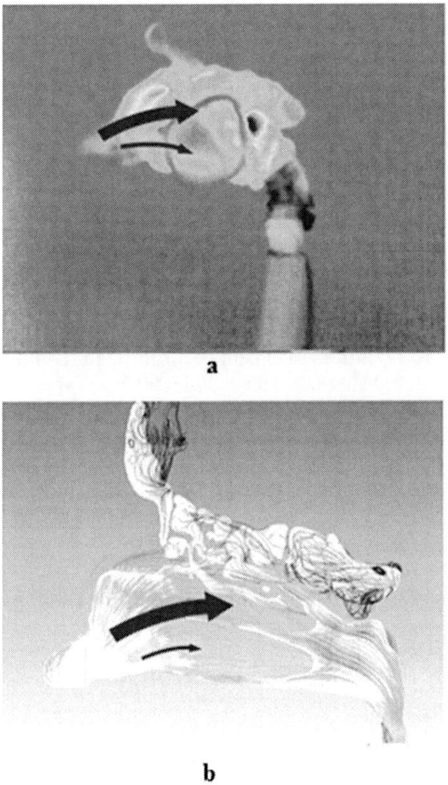

Fig. 13. Results of (a) thermography and (b) numerical modelling of a platyform nasal cavity

References

[1] B. M. Sagalovich, *Physiology and Pathophysiology of the Upper Respiratory Tract* Moscow: Meditsina. 1964.

[2] G. Mlynski, S. Grutzenmacher, S. Plontke, B. Mlynski, and C. Lang, "Correlation of nasal morphology and respiratory function", *J. Rhinology*, 2001, 39, 197-201.

[3] G. Mlynski, S. Grutzenmacher, S. Plontke, W. Grutzmacher, B. Mlynski, P, C. Lang "A method for studying nasal airflow by means of fluid dynamics experiments", J. z. Med. Phy., 2000, 10, 207-214.

[4] K. Keyhani, P. W. Scherer, and M. M. Mozell, "Numerical simulation of airflow in the human nasal cavity", *J. Biomech. Eng.*, 1995, 117, 429.

[5] J. Wen, K. Inthavong, Z. F. Tian, J. Y. Tu, C. L. Xue and C. G. Li, "Airflow Patterns in Both Sides of a Realistic Human Nasal Cavity for Laminar and Turbulent Conditions", *in Proceedings of the 16th Australasian Fluid Mechanics Conference*, Crown Plaza, Gold Coast, Australia, 2007, pp. 68–74.

[6] K. Zhao, P. Dalton, G. C. Yang, and P. W. Scherer, "Numerical modeling of turbulent and laminar airflow and odorant transport during sniffing in the human and rat nose", *J. Chem. Senses,* 2006, 31 (2), 107-118.

[7] J. Pennecot, M. Ruetten, R. Kessler, and C. Wagner, "CFD Mesh of the Human Nose and Flow Simulation", *in Proceedings of the 8th World Congress on Computational Mechsnics*, Venice, 2008.

[8] G. Lukyanov, A. Rassadina, V. Usachev, "Comparison and the analysis of the processes of the movement of air through the human breathing system and its natural model", 2005, *2005 International Conference on Physics and Control, PhysCon 2005, Proceedings*

[9] A. G. Malyshev, N. K. Zhumashev, G. N., Lukyanov, K. D. Mynbaev, A. A. Rassadina, "Application of a LED-photodiode optocouple for the study of human respiratory function", *Journal of Physics: Conference Series*, 2015

[10] Lukyanov G. N., Rassadina A. A., Makarov S. L. "A new method and a sensor for diagnosis of the respiratory diseases", *2nd International Scientific Symposium "Sense. Enable. SPITSE"* (St. Petersburg, 22 June-03 July 2015) 2015, pp. 219-222

[11] J. Znu *A numerical study of airflow through human upper airways*. Singapore: National University of Singapore. 2012. 192 p.

[12] J. H. Zhu, H. P. Lee, K. M. Lim, Sh. J. Lee, D. Y. Wang, "Evaluation and comparison of nasal airway flow patterns among three subjects from Caucasian, Chinese and Indian ethnic groups using computational fluid dynamics simulation", *J. Respiratory Physiology & Neurobiology*, 175 (2011), 62-69

[13] J. A. Keeler, A. Patki, C. R. Woodard, D. O. Frank-Ito, "A Computational Study of Nasal Spray Deposition Pattern in Four Ethnic Groups", *J. Aerosol Med Pulm. Drug Deliv*. 2016 Apr; 2 9(2): 153-66. doi: 10.1089/jamp.2014.1205.

[14] K. Zhao, J. Jiang, "What is normal nasal airflow? A computational study of 22 healthy adults", *J. Int. Forum Allergy Rhinol*. 2014 Jun; 4(6):435-46. doi: 10.1002/alr.21319

[15] X. B. Chen, H. P. Lee, V. F. Chong, Y. Wang de, "Impact of inferior turbinate hypertrophy on the aerodynamic pattern and physiological functions of the turbulent airflow: a CFD simulation model", *J. Rhinology*. 2010 Jun; 48(2): 163-8. doi: 10.4193/Rhin09.093.

[16] R. V. Neronov, G. N. Luk'yanov, A. A. Rassadina, A. A. Voronin, A. G. Malyshev, "The effect of the nasal cavity form on airflow distribution during inhalation", *J. Russian Otorhinolaryngology*, 2017, 1 (86), 83-94

G. N. Lukyanov, A. A. Voronin, A. A. Rassadina, "Simulation of convective flows in irregular channels on the example of the human nasal cavity and paranasal sinuses" *J. Technical Physics. The Russian Journal of Applied Physics*. 2017, 62 (3), 484-489.

[17] Grassberger P., Procaccia I. Characterization of strange attractors//*Phys. Rev. Lett*. 1983. N 50. pp. 346-349.

[18] A. Voronin, G. Luk'yanov, E. Frolov "Detached-eddy simulation of turbulent airflow", *Scientific and Technical Journal of Information Technologies, Mechanics and Optics*, 2014, №1 (89) 187-192

[19] Wilcox D.C. Turbulence Modelling for CFD. DCW Industries, Inc., 1993. 460 p.

[20] Menter F. R. Zonal Two Equation *k–ω* Turbulence Models for Aerodynamic Flows // *AIAA Paper*. 1993. V. 93 (2906). 21 p.

CHAPTER 4.2

TEXTURAL FEATURES FROM BASELINE CT AND BLOOD FLOW MAPS ASSOCIATED WITH OVERALL SURVIVAL IN NON-SMALL CELL LUNG CANCER

SERENA BAIOCCO*, DOMENICO BARONE†, GIAMPAOLO GAVELLI† AND ALESSANDRO BEVILACQUA*

*University of Bologna, Bologna, Italy
Email: {s.baiocco, alessandro.bevilacqua}@unibo.it
†IRCCS - Istituto Scientifico Romagnolo per lo Studio e la Cura dei Tumori (IRST), Meldola (FC), Italy Email: Email: {d.barone, g.gavelli}@irst.emr.it

Abstract. Tumour heterogeneity makes the practice of clinical oncology very challenging, as it leads a great variability in tumour response to therapy. This phenomenon highlights the complex biology of the micro-environment of the tumour, characterised by both spatial and temporal variability; this is increased by the presence of chaotic blood vessels within tumour tissue. This all translates into variations in tissue morphology, angiogenesis, metabolism, cellular proliferation and metastatic potential. For this reason, over the past decade, quantifying the salient features of intra-tumoural heterogeneity has gained significant attention, leading to a newfound emphasis on texture analysis. Computed tomography (CT) has always been considered a reference technology for the morphological analysis of organs and tissues, permitting the capture of *in vivo* spatial heterogeneity of tumours. The need to detect haemodynamic tumour features has stimulated the use of CT perfusion (CTp), a promising functional imaging technique in the oncological field. CTp allows detection of the presence of abnormal haemodynamic patterns in a tumour by analysing the temporal variations in tissue that occur following intravenous administration of a contrast medium.

This work presents the extraction of meaningful statistical and textural features from unenhanced CT images and perfusion maps. These may be used as prognostic image-based biomarkers for lung tumours.

I. Introduction

Non-small cell lung cancer (NSCLC) is the leading cause of cancer mortality worldwide [1]. Despite the introduction of novel advanced therapies, NSCLC remains associated with poor prognoses and low overall survival (OS) [2]. Tumour staging is probably the most important prognostic parameter for survival, though different outcomes within the same stage grouping suggest that other factors also influence survival [3]. One of the challenges in this age of personalized medicine remains the identification of more effective prognostic markers for better stratification of NSCLC patients [4] and has strong implications for patient management, particularly as regards treatment choices [5].

NSCLC tissues are highly heterogeneous at different scales, reflecting the presence of angiogenesis, hypoxia, high cell density, necrosis, and haemorrhage [6], [7]. Angiogenesis is the process of chaotic sprouting of blood vessels from pre-existing ones [8], [9]. With the enlargement of vascularization, tumour tissue is able to receive nutrients and create paths for cells to leave or enter the blood's circulation. This process leads to tumour growth associated with increased interstitial pressure; this may lead to local reductions in blood flow (BF) and, consequently, to inadequate oxygen delivery to cells. This condition is known as hypoxia and has a decisive role in tumour response to treatment [10].

By its very nature, quantitative tumour evaluation through molecular characterisation, involving biopsies or invasive surgery to acquire small tissue specimens, cannot adequately represent the features of the tumour as a whole [11]. On the contrary, imaging is able to provide voxel-based information for the whole tumour [12]. It has great potential to capture *in vivo* intra-tumoural heterogeneity non-invasively [13], paving the way for routine tumour monitoring and therapy response evaluation.

Primarily, the way images are clinically interpreted is visual [14]. The aim of the clinician is to detect potential abnormalities, recognise patterns, and link the patterns perceived to possible diagnoses. Therefore, the success of image interpretation strongly depends on the skills of the clinician [15]. Consequently, a qualitative evaluation of heterogeneity will be prone to subjective variability. As a result, despite tissue heterogeneity proven to be strongly related to tumour aggressiveness [16], prognosis [17], and survival [18], its quantitative evaluation is still not included in clinical practice.

Over the past decade, several studies have presented an analysis of tumour heterogeneity so as to identify more accurate prognostic biomarkers through texture analysis [14], [19], [20]. Texture analysis covers several different techniques. These are not new to the medical field, but date back to the early 1970s, when texture measures were first suggested for the detection of lung disease [21]. Their major advantage in clinical practice is that these mathematical approaches can offer information that is not visible to the human eye in an objective way, exploiting routinely acquired data [22].

Recently, CTp has come to be accepted as a clinical technique in oncology, primarily for the early evaluation of tumour response to anti-angiogenic therapies (i.e. therapies seeking to reduce tumour vasculature) [23]. Indeed, CTp allows the capture of vascular patterns in tumours through the analysis of time-concentration curves (TCCs), representing temporal variations in tissue density [24], which are directly proportional to the quantity of contrast medium within the tissue. One of the most significant perfusion parameters is the BF, which permits to detect the earliest functional changes on tumour vasculature even within the first week of anti-angiogenic therapy [25]. Moreover, this parameter has proved a high correlation with tissue biomarker micro-vessel density (MVD) [26], [27].

Analysis of BF maps could allow the detection of abnormal vascular patterns within tumour tissue. However, several factors may affect the reliability of BF maps. Of these, cardiac and respiratory motion artefacts, physics-based artefacts, and partial volume artefacts may all alter the calculation of perfusion parameters. Therefore, a proper denoising procedure is needed to characterise the haemodynamic heterogeneity and, finally, to derive quantitative features [28].

In this work, we analyse the intra-tumoural heterogeneity of NSCLC emerging from unenhanced CT images and BF maps, at the diagnosis stage. To this end, BF maps were computed using the maximum-slope method. A proper reliability analysis of the BF maps was performed to automatically detect and remove those pixels and regions suffering from high computing errors due to artefacts. The correlations between the OS parameter and statistical feature-pairs computed with CT images and BF maps were explored to assess whether these features could work as prognostic image-based biomarkers for NSCLC. The discriminative ability of the feature-pair was analysed in comparison with the stage of the tumour, which is commonly associated with the OS parameter. The features derived from functional BF maps showed high reproducibility and a strong correlation with the OS parameter, being more effective than those directly derived from CT images; this suggests a promising prognostic clinical application of CTp. The features extracted from non-contrast enhanced CT images, easily derivable from data routinely acquired in clinical practice, have been proved to be highly reproducible and more effective than the stage parameter.

II. Materials and Methods

A. Study population

As reported in a previous work [29], this study was approved by the Institutional Review Board, waiving informed consent for the retrospective data analysis of the patients. 36 consecutive patients (25 men, 11 women; age range 36-81 years) with primary NSCLC, subdivided into 28 adenocarcinoma (AC), 6 squamous-cell carcinoma (SCC), and 2 large-cell carcinoma (LCC), were enrolled for data analysis between September 2010 and December 2012. According to the TNM classification (seventh edition) of malignant tumours, 2 patients were diagnosed as stage IB; 3 patients as IIIA; 4 patients as IIIB; and 27 patients as tumour stage IV. Patient inclusion criteria included: (a) over eighteen years old; (b) with histologically verified NSCLC; and (c) no prior history of chemotherapy, surgery, or thoracic radiation therapy.

Exclusion criteria included: (a) clinically significant cardiovascular disease; (b) pregnancy or lactation; (c) a known history of deep vein thrombosis or pulmonary embolism; (d) patients lost to follow-up for unknown reasons (n = 8); (e) patients alive at the time of the study (n = 4); (f) not having a long axial diameter of a lesion greater than 10 mm in at least one slice (n = 3); and (g) examination hardly affected by physics-based artefacts (n = 2).

Finally, 19 patients were considered for analysis. TNM staging was considered as the reference point to predict survival. This variable was dichotomised so as to divide the subjects in early stage patients (stage I–IIIA), likely receiving curative surgery or curative chemo-radiotherapy; and advanced stage patients (stage IIIB–IV), likely undergoing non-curative chemotherapy, radiotherapy, or molecular therapies [29].

Table 1 reports the above information and also includes patients' histological subtype (ACC, SCC, LCC) and OS data, defined as the time interval in months between the date of the baseline CT examination and the date of death.

Table I. Summary of Histological Diagnosis, Tumour Stage, and OS Data (Months) for Each Patient.

Patient ID	OS	Diagnosis	Stage
ID1	6	LCC	IIIB
ID2	4	AC	IV
ID3	14	SCC	IB
ID4	13	AC	IV
ID5	5	AC	IV
ID6	5	AC	IB
ID7	6	AC	IV
ID8	10	AC	IV
ID9	17	SCC	IV
ID10	52	SCC	IV
ID11	6	AC	IV
ID12	6	AC	IV
ID13	4	AC	IV
ID14	8	AC	IV
ID15	12	AC	IV
ID16	11	SCC	IIIB
ID17	12	AC	IV
ID18	0	AC	IV
ID19	7	AC	IV

B. CTp protocol

At diagnosis, patients underwent axial CTp, which was performed on a 256-slice CT system (Brilliance iCT, Philips Medical Systems, Best, Netherlands) and laid in the supine (feet first) position. An initial unenhanced low-dose full-body axial CT scan was performed to identify the target lesion at baseline condition. Following this, an intravenous 50 mL bolus of contrast agent (Iomeron, Bracco, Milan, Italy) was injected at 5 mL/s for axial cine contrast enhanced CT. Five seconds later, a CTp scan of 25-second duration was performed for each patient under the breath-hold condition. This protocol yielded 20 scans, each corresponding to different sampling instants, with 55 mm of z-coverage (11 slices × 5-mm slice thickness, 0.4-sec rotation time, at 80 kV, 250 mA). Image data were reconstructed into 220 cine images (512 × 512 pixel, 11 slices, 350 mm × 350-mm, 5-mm slice spacing, 1.25-sec temporal resolution).

C. Generation of perfusion maps

The lesions and arterial input were selected with the consensus of two radiologists (D. B. and G. G., > 25 years' experience each). For each lesion, the radiologists manually drew the region of interest (ROI) on a reference slice and then performed a 3D rigid alignment using the following procedure. The lesion ROI was translated onto the slices of the reference sequence, so that the contour of the lesion visually matched. When opportune, adjacent slice levels were chosen by the radiologists to achieve the best match with the reference slice [30] and, consequently, the "best" sequence (Fig. 1).

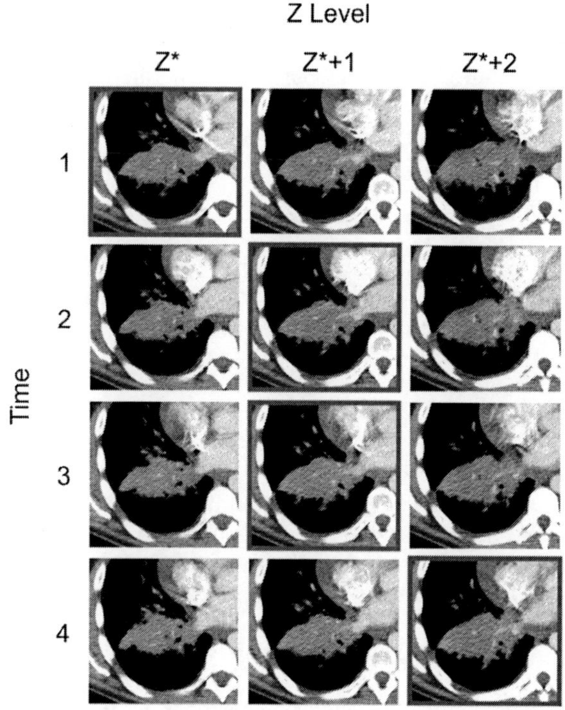

Fig. 1. Lung tumour perfusion. Example of the selection of the best sequence to achieve the best matching with the Z levels contiguous to the reference slice level, represented by Z*

BF values were computed by considering the first-pass phase [31] and modelling each voxel as a single-compartment including both intravascular and extra-vascular spaces. This permitted the performance of short-time examinations, reducing motion artefact effects since the patients were able to stay in apnoea. The model adopted is characterised by a single inlet, the arterial input, with an arterial blood plasma temporal concentration $C_A(t)$, and a single outlet, the venous outlet, with a venous blood plasma temporal concentration $C_V(t)$. $C_T(t)$ represents the TCC of the tissue. The relation between $C_A(t)$, $C_V(t)$ and $C_T(t)$ can be expressed through the Fick principle, representing the conservation of mass:

$$\frac{dC_T(t)}{dt} = \frac{BF}{V_T}[C_A(t) - C_V(t)] \tag{1}$$

The assumption of no venous outflow during the measurement period $C_V(t) \approx 0$ (i.e., no contrast agent has reached the venous side of the circulation yet) leads to the following approximation:

$$\frac{dC_T(t)}{dt} = \frac{BF}{V_T}[C_A(t) - C_V(t)] \approx \frac{BF}{V_T}C_A(t) \tag{2}$$

This implies that the slope of the tissue curve, $dC_T(t)/dt$, reaches a maximum (steepest slope) when the TCC of the arterial input, $C_A(t)$, reaches its peak density [32]. Thus, the BF, expressed in mL/min/100 g, is given by:

$$\frac{BF}{V_T} \approx \frac{\left[\frac{dC_T(t)}{dt}\right]_{max}}{[C_A(t)]_{max}} \tag{3}$$

This is the single-compartment formulation, also known as the Mullani-Gould model. The simplification adopted may not hold for organs characterised by complex microcirculation, such as the spleen and kidneys [33], but it is adequate for most clinical applications [34].

The BF values of the reference slice are represented using functional colorimetric maps, as shown in Fig. 2. Blue regions highlight low perfusion values, while regions characterised by the "hottest" colors emphasise the areas of highest perfusion. These maps provide a useful view of the perfusion and the functional heterogeneity of the tumour, highlighting for instance the presence of hypodense regions, haemorrhages, blood vessels, and bronchi.

In order to obtain the TCC signal for each voxel, $y_{TCC}(t)$, we adopted a sigmoid-shape model, as reported in Fig. 3. This sigmoid is modelled through the Hill equation in Eq. (4) and is able to robustly fit the main trend of the concentration samples [35]:

$$y_{TCC}(t) = E_0 + (E_{max} - E_0)\frac{t^{\alpha}}{(EC_{50}+t)^{\alpha}} \quad (4)$$

where E_0 is the baseline concentration of the tissue, i.e. before the arrival of the tracer; E_{max} is the saturation value of the concentration reached after its arrival; EC_{50} is the time instant at the half-maximum value of the concentration; and α is the parameter that mainly affects the slope of the sigmoid curve. Curve fitting is achieved using an in-house fitting algorithm based on the nonlinear, least squares, Levenberg-Marquardt minimization algorithm (lsqcurvefit, Matlab©; MathWorks, Natick, MA, USA).

D. *Denoising perfusion maps*

Denoising is a necessary step before feature extraction. Essentially, it consists in detecting and excluding unreliable BF values. BF values strictly lower than 1 mL/min/100 g were automatically excluded, being considered improbable as physiological values and more likely attributable to numerical errors, as the perfusion computation algorithm forces BF values to be positive.

For each TCC signal, the quality of the fitting was assessed by considering the residuals, ε, computed as the difference between the observed samples, Y_{TCC}, and the relative values obtained through the fitting, y_{TCC}, for each time instant t. ε is defined as in Eq. (5):

$$\varepsilon(t) = |Y_{TCC}(t) - y_{TCC}(t)| \quad (5)$$

As a goodness-of-fit index for the single TCC, the temporal mean value, με, of residual ε relative to the considered pixel was computed. The distribution of this index was used to detect BF values characterised by high fitting errors in order to exclude them from the analysis, since they are computed on TCCs that are incorrectly fitted.

High errors are usually associated with the presence of noise, artefacts, and anatomical structures, such as bony regions, blood vessels, and bronchi. Using this index, these structures can be automatically excluded. As shown in Fig. 2, at the end of this automatic denoising process, the removed BF values appear highlighted in pink in the colorimetric maps. As one can see, the denoising method also removes unreliable regions as borders, which are affected by the partial volume effect, while still keeping a wide range of BF values [36].

E. Texture analysis

Feature generation and selection were performed using in-house software based on Matlab© (MathWorks, Natick, MA, USA) [37]. First (histogram) and second-order (grey-level co-occurrence matrix) statistical features were computed for reference slices on unenhanced CT images and BF maps.

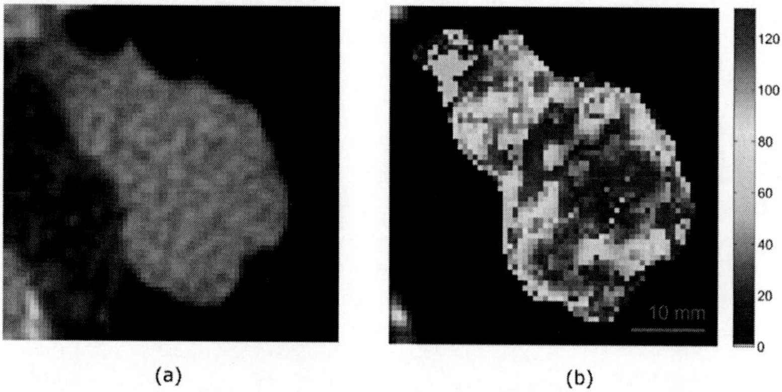

Fig. 2. Unenhanced CT image (a) and corresponding BF map (b) of a lung tumour (squamous-cell carcinoma) [30]

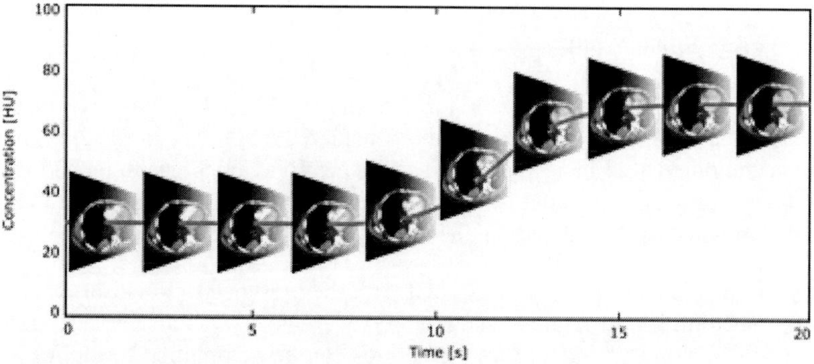

Fig. 3. Representation of the sigmoid model used for the fitting procedure of the voxel values, belonging to the tumour reference slice [30]

First-order statistics are based on the histogram of pixel intensity values in the image. The features derived from value distribution depend only on single pixel values and not on the interaction or co-occurrence of neighbouring pixel values. They are global features, which describe how intensity values within a ROI are distributed. Let $x = \{x_1, x_2,..., x_N\}$ be the set of L values of the N voxels in the image. Let $h = \{h_1, h_2,.., h_L\}$ be the histogram with count n_i of each intensity level i in x. The occurrence probability, p_i, for each intensity level is $p_i = \frac{h_i}{N}$. Accordingly, we can use the following definitions:

Mean μ

$$\mu = \frac{1}{N}\sum_{k=1}^{N} x_k \tag{6}$$

Standard deviation σ

$$\sigma = \left(\frac{1}{N}\sum_{k=1}^{N}(x_k - \mu)^2\right)^{\frac{1}{2}} \tag{7}$$

Skewness S

$$S = \frac{\frac{1}{N}\sum_{k=1}^{N}(x_k - \mu)^3}{\left[\frac{1}{N}\sum_{k=1}^{N}(x_k - \mu)\right]^{\frac{3}{2}}} \tag{8}$$

Kurtosis K

$$K = \frac{\frac{1}{N}\sum_{k=1}^{N}(x_k - \mu)^4}{\left[\frac{1}{N}\sum_{k=1}^{N}(x_k - \mu)^2\right]^2} - 3 \tag{9}$$

Coefficient of variation CV

$$CV = \frac{\sigma}{\mu} \tag{10}$$

Entropy E, also known as Shannon entropy

$$E = -\sum_{i=1}^{L} p_i \log_2 p_i \tag{11}$$

Uniformity U

$$U = \sum_{i=1}^{L} {p_i}^2 \tag{12}$$

Other parameters derived from the histogram analysis include: the median; the intensity level dividing a distribution into two halves; and the minimum and the maximum, that is, the lowest and highest pixel values.

Second-order statistics are based on the likelihood of observing pixel pairs of specific intensity values. These features belong to the textural features, since they can represent the spatial distribution of image values. More specifically, these features are based on the joint probabilities and provide co-occurrence measurements performed on matrices, known as grey level co-occurrence matrices (GLCMs), which reflect the spatial grey-level dependence. Actually, these matrices may contain any type of entity, besides grey levels. Each cell of these matrices represents how often a couple of pixel values (i, j) occurs, this being the basis for measuring the relationship between them. Therefore, these matrices provide valuable information about images, since, as reported by Haralick et al., tone and texture, which are always present in the images, bear an inextricable relationship [38].

The features extracted from the GLCM provide statistical information of the image regarding the distribution of pixel pairs along one of the image directions. GLCM provides a new image representation [39] and is an estimate of second-order joint probability, pij, of the intensity values of two pixels, i and j, separated by a distance δ ($\delta = 1$ for neighbouring pixels) along a given direction θ (where θ can usually be horizontal, vertical, diagonal, and anti-diagonal, i.e. $\theta = \{0°, 45°, 90°, \text{and } 135°\}$). This joint probability can be represented with a square matrix, pij, with row and column dimensions equal to the number of discrete intensity levels, L, in the image. Each pij element contains the frequency of a combination of intensity level pairs, i and j, at a certain distance δ, along a specific direction θ. The resulting GLCM is a symmetrical matrix, as shown in Fig. 4. For images tending to be flat, i.e. uniform with no texture, the resulting GLCM would be almost diagonal. By contrast, for images with increased local intensity variations, i.e. with increased texture, the resulting GLCM will be characterised by more off-diagonal values. The features that will be derived from GLCM are local features that should encode, in an efficient way and useful for classification tasks, the relevant information present in GLCM structures. Some GLCM features reported below have a clear correspondence with human texture perception (e.g., coarseness, smoothness), while others do not represent any specific visual property, but still encode textural information with a high discriminatory power [40].

0	0	0	1
1	1	1	1
2	2	2	3
3	3	4	4

(a)

	0	1	2	3	4
0	4	1	0	0	0
1	1	6	0	0	0
2	0	0	4	1	0
3	0	0	1	2	1
4	0	0	0	1	2

$\delta = 1, \theta = 0°$

(b)

Fig. 4. Small image (a) and relative co-occurrence matrix (b) for $\delta = 1$ and $\theta = 0°$. The joint probability is obtained by dividing the matrix in (b) for the total number of possible pixel pairs [30]

Joint maximum, M_J, is the probability corresponding to the most common GLCM co-occurrence and is defined as

$$M_J = \max(p_{ij}) \tag{13}$$

Joint average, μ_J, is the weighted sum of the joint probabilities and is defined as

$$\mu_J = \sum_{i=1}^{L}\sum_{j=1}^{L} i\, p_{ij} \tag{14}$$

Joint variance, σ_J^2, is a measure of the variability of GLCM co-occurrences and is defined as:

$$\sigma_J^2 = \sum_{i=1}^{L}\sum_{j=1}^{L}(i - \mu)^2\, p_{ij} \tag{15}$$

where μ = joint average.

Second-order entropy, E_J, is defined as

$$E_J = -\sum_{i=1}^{L}\sum_{j=1}^{L} p_{ij}\, log_2 p_{ij} \tag{16}$$

Angular second moment, *ASM* [38], also called energy and uniformity, is defined as

$$ASM = \sum_{i=1}^{L}\sum_{j=1}^{L} p_{ij}^2 \tag{17}$$

Contrast, *C*, a measure of the intensity levels variations, is defined as

$$C = \sum_{i=1}^{L} \sum_{j=1}^{L} (i-j)^2 \, p_{ij} \tag{18}$$

Inverse difference, *IDF*, is defined as

$$IDF = \sum_{i=1}^{L} \sum_{j=1}^{L} \frac{p_{ij}}{1+|i-j|} \tag{19}$$

Normalised inverse difference, *IDFN*, is defined as

$$IDFN = \sum_{i=1}^{L} \sum_{j=1}^{L} \frac{p_{ij}}{1+|i-j|/L} \tag{20}$$

Inverse difference moment, *IDM*, a measure of local homogeneity, is defined as

$$IDM = \sum_{i=1}^{L} \sum_{j=1}^{L} \frac{p_{ij}}{1+(i-j)^2} \tag{21}$$

Normalised homogeneity, *HN*, is defined as

$$HN = \sum_{i=1}^{L} \sum_{j=1}^{L} \frac{p_{ij}}{1+(i-j)^2/L^2} \tag{22}$$

Correlation, *CORR* [38], is defined as

$$CORR = \frac{\sum_{i=1}^{L} \sum_{j=1}^{L} ij \, p_{ij} - \mu_x \mu_y}{\sigma_x \sigma_y} \tag{23}$$

where

$$\mu_x = \sum_{i=1}^{L} i \sum_{j=1}^{L} p_{ij} \tag{24}$$

$$\mu_y = \sum_{j=1}^{L} j \sum_{i=1}^{L} p_{ij} \tag{25}$$

$$\sigma_x = \sum_{i=1}^{L} (i - \mu_x)^2 \sum_{j=1}^{L} p_{ij} \tag{26}$$

$$\sigma_y = \sum_{j=1}^{L} (j - \mu_y)^2 \sum_{i=1}^{L} p_{ij} \tag{27}$$

Autocorrelation, *A*, is defined as

$$A = \sum_{i=1}^{L} \sum_{j=1}^{L} i \, j \, p_{ij} \tag{28}$$

F. Statistical Analysis

The principal purpose of this work was to determine whether statistical and textural features directly extracted from non-contrast enhanced CT images and denoised BF maps could work as surrogate prognostic biomarkers. Given the sample size, to avoid overfitting, no more than two features were analysed jointly. Feature selection was performed through a k-means ($k = 2$) clustering algorithm, using the squared Euclidean distance as a measure of similarity. Following this, to identify the most promising prognostic feature-pairs, the correlation with the OS parameters was analysed automatically. Mean survival for the entire cohort of patients ($\mu_{OS} = 10.5$) was used as a cut-off to separate patients with the highest (OS $\geq \mu_{OS}$) and the lowest (OS $< \mu_{OS}$) survival expectation. In order to determine to what extent the feature means differ between the groups identified by the clustering algorithm, a multivariate analysis of variance (MANOVA) was performed (p-value < 0.001). Sensitivity (SE), specificity (SP), positive predictive value (PPV), negative predictive value (NPV), and accuracy (ACC) were calculated to quantify the discriminatory ability of both selected features and stage, i.e. the variable used as reference for survival prediction. In particular, a low survival expectation and a high survival expectation were defined as condition true and condition false, respectively. An advanced stage condition is considered a positive prediction, while an early stage is considered a negative one. Consequently, patients at an advanced stage and with a low survival expectation are true positives (TP) and patients at an early stage and with a high survival expectation are true negatives (TN). On the other hand, patients at an advanced stage and with a high survival expectation are considered to be false positives (FP), while patients at an early stage and with a low survival expectation are false negatives (FN) [30].

In order to assess the potential impact that inter-observer variability and the partial volume effect might have on reproducibility, automatic erosion was computed to alter the lesion boundary by one pixel. Intra-class correlation coefficients (ICCs) were calculated by considering the feature values derived from the eroded ROIs.

Statistical analysis was performed using Matlab© (MathWorks, Natick, MA, USA).

III. Results

Feature-pairs derived from BF maps can be characterised as having the greatest discriminatory power in distinguishing patients with different survival expectations. More specifically, first-order entropy (*E*) and the inverse difference moment (*IDM*), derived from the BF maps, were highly effective at separating patients with different survival expectations (SE = 86 %, SP = 75 %, PPV = 68 %, NPV = 90 %, ACC = 79 %). This feature-pair also gave the best pair correlation with the OS parameter of those derived from unenhanced CT images (SE = 67 %, SP = 71 %, PPV = 80 %, NPV = 56 %, ACC = 68 %). In addition, among the feature-pairs derived from the BF maps, comparable performance can be obtained with first-order entropy and second-order uniformity (*ASM*) (SE = 86 %, SP = 75 %, PPV = 68 %, NPV = 90 %, ACC = 79 %).

E and *IDM*, as well as *E* and *ASM*, computed on BF maps are able to group patients with the lowest (OS < μ_{OS}) and highest (OS ≥ μ_{OS}) survival expectation. Fig. 5 graphically shows the distribution of the feature values extracted from BF maps for low and high survival expectation. As one can see, high *E* and low *IDM* or low *ASM* highlighted patients with the worst prognosis (OS < μ_{OS}), while low *E* and high *IDM* or high *ASM* characterised patients with the best prognosis in this study (OS ≥ μ_{OS}). In Fig. 6, box plots of *E* and *IDM* derived from unenhanced CT images for low (OS < μ_{OS}) and high survival expectation (OS ≥ μ_{OS}). As one can see, analogously to the most promising BF-based features, high *E* and low *IDM* categorise patients with the worst prognosis (OS < μ_{OS}), while low *E* and high *IDM* characterise those with the best prognosis (OS ≥ μ_{OS}).

The outcome of MANOVA proves that both the BF-based feature-pairs, *E* and *IDM*, as well as *E* and *ASM*, differ significantly for the two groups of patients, with p-values of 6.4 × 10−5 and 1.5 × 10−4, respectively. Furthermore, the CT-based feature-pair, *E* and *IDM*, differs significantly for the two groups of patients, with *p*-values of 9.8 × 10−6.

TNM staging (SE = 92 %, SP = 14 %, PPV = 65 %, NPV = 50 %, ACC = 63 %) performed more poorly as a survival predictor.

As far as the reproducibility analysis was concerned, computed erosion altered the tumour ROI by 11.1 ± 4.3 %. Reproducibility errors for *E*, *IDM*, *ASM* derived from BF maps were only 0.8 ± 0.7 %, 3.0 ± 2.6 % and 6.9 ± 6.3 %, respectively. Analogously, reproducibility errors for *E* and *IDM* derived from baseline CT images were 0.7 ± 0.7 % and 2.8 ± 2.0 %, respectively. The ICC for *E*, *IDM* and *ASM* derived from BF maps were 0.99, 0.99 and 0.97, respectively; while ICC for *E* and *IDM* derived from baseline CT images were 0.99 and 0.99, indicating that the selected features were highly reproducible.

IV. Conclusion

Although only essential parameters are routinely quantified in clinical practice, this study has shown that statistical and textural features, to capture intra-tumoural heterogeneity, could work as a prognostic image-based biomarker for NSCLC. In particular, two feature-pairs derived from denoised BF map were strongly associated with the OS parameter. As expected, these features, being derived from BF maps, carry significant functional information content, which mainly reflects the presence of hypoxic regions and vascular abnormalities. Therefore, this finding represents a promising approach for the utilization of CTp in clinical practice.

Despite being less promising than BF-based features, it is worth noting that a feature-pair, derived from unenhanced CT images, was more specific than TNM staging in discriminating patients with different survival expectations. It is also noticeable that this feature-pair gives similarly promising results for BF maps, highlighting that the intra-tumoural heterogeneity in NSCLC, the phenomenon investigated here, is independently detectable in the modality adopted. This is an encouraging outcome since these features may be easily derived from conventional CT examinations that, contrarily to CTp examinations, are routinely acquired in the clinical practice.

The results of this study coherently showed that greater intra-tumoural heterogeneity was associated with more aggressive tumours and is strongly reflected in the OS parameter. It is worth noting that global entropy (*E*) gave promising results if derived from BF maps or from unenhanced CT images. This is not really surprising as *E* is a meaningful feature representing value irregularity and better known as a measure of information content. High entropy correlated to a poorer prognosis. Analogously, local homogeneity and uniformity correlated to a better prognosis.

The encouraging outcomes achieved in this work should give impetus to further study involving larger patient cohorts, as well as analyses of repeatability and reproducibility. The results also encourage the deepening of the unknown relationship between the results of imaging and biological features. This would permit a comprehensive understanding of the specific cancer biology of an individual patient and infer phenotypic signatures containing predictive or prognostic values. Unlike biopsies, quantitative imaging could permit the longitudinal monitoring of intra-tumour heterogeneity and, consequently, of "specific" cell habitats and their change over time, since clinical examination can be easily repeated during the therapy [41].

Quantitative image analysis has the potential to improve risk stratification in NSCLC, providing useful information for disease evaluation and management, and guiding clinicians in the choice of more optimized, patient-specific treatments. This has a fundamental role in the treatment of NSCLC patients, who are usually high-risk candidates for invasive procedures and surgical intervention [42].

In conclusion, analysis of tumour heterogeneity performed with imaging provides valuable information regarding *in vivo* tumour characteristics, which should be further explored to be translated into precision radiation oncology and the personalisation of medicine.

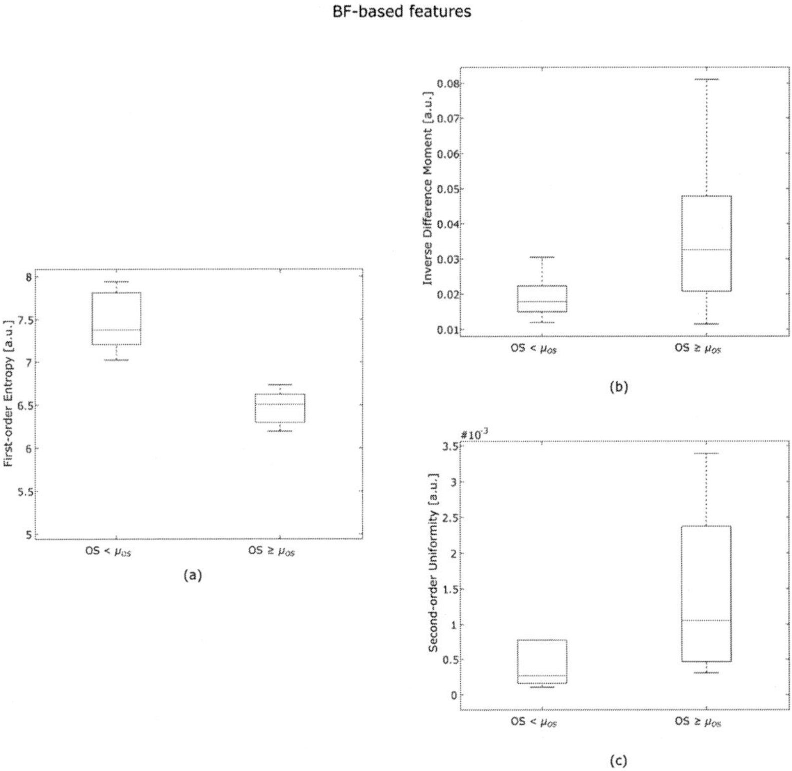

Fig. 5. Box plots showing first-order entropy (a); inverse difference moment (b); and second-order uniformity derived from BF maps for $OS < \mu_{OS}$ (left) and $OS \geq \mu_{OS}$ (right). The median is indicated as a red line in the boxes, while the vertical box size gives the interquartile range [30]

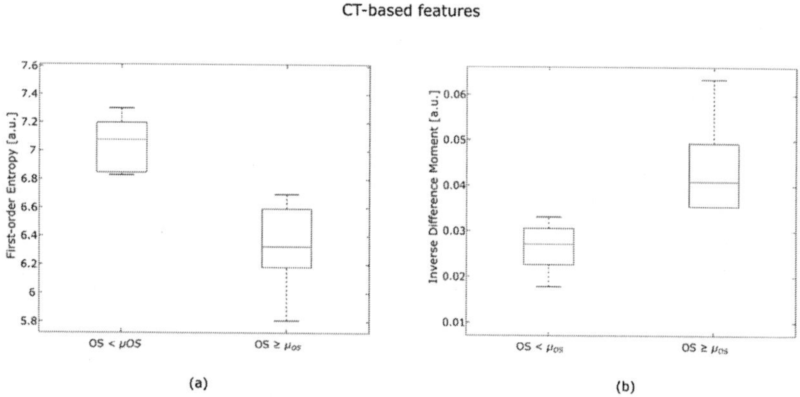

Fig. 6. Box plots showing first-order entropy (a); and inverse difference moment (b) derived from unenhanced CT images for $OS < \mu_{OS}$ (left) and $OS \geq \mu_{OS}$ (right). The median is indicated as a red line in the boxes, while the vertical box size gives the interquartile range [30]

References

[1] Zhao Chen, Christine M Fillmore, Peter S Hammerman, Carla F Kim, and Kwok-Kin Wong. Non-small-cell lung cancers: a heterogeneous set of diseases. *Nat. Rev. Cancer*, 14(8):535–546, 2014.

[2] Martin Reck, David F Heigener, Tony Mok, Jean-Charles Soria, and Klaus F Rabe. Management of non-small-cell lung cancer: recent developments. *Lancet*, 382(9893):709–719, 2013.

[3] Marliese Alexander, Rory Wolfe, David Ball, Matthew Conron, Robert G Stirling, Benjamin Solomon, Michael MacManus, Ann Officer, Sameer Karnam, Kate Burbury et al. Lung cancer prognostic index: a risk score to predict overall survival after the diagnosis of non-small-cell lung cancer. *British journal of cancer*, 117(5):744, 2017.

[4] Balaji Ganeshan, Elleny Panayiotou, Kate Burnand, Sabina Dizdarevic, and Ken Miles. Tumour heterogeneity in non-small cell lung carcinoma assessed by CT texture analysis: a potential marker of survival. *Eur. Radiol.*, 22(4):796–802, 2012.

[5] Stephen G Spiro, Nichole T Tanner, Gerard A Silvestri, Sam M Janes, Eric Lim, Johan F Vansteenkiste, and Robert Pirker. Lung cancer: progress in diagnosis, staging and therapy. Respirology, 15(1):44–50, 2010.

[6] Fergus Davnall, Connie SP Yip, Gunnar Ljungqvist, Mariyah Selmi, Francesca Ng, Bal Sanghera, Balaji Ganeshan, Kenneth A. Miles, Gary J. Cook, and Vicky Goh. Assessment of tumour heterogeneity: an emerging imaging tool for clinical practice? Insights into imaging, 3(6):573–589, 2012.

[7] Alessandro Bevilacqua and Serena Baiocco. Automatic classification of lung tumour heterogeneity according to a visual-based score system in dynamic contrast enhanced CT sequences. Int. J. Mod. Phys. C, pages 1650106–1–1650106–14, 2016.

[8] William D Figg and Judah Folkman. Angiogenesis: an integrative approach from science to medicine. Springer Science & Business Media, 2008.

[9] Gordon C Jayson, Robert Kerbel, Lee M Ellis, and Adrian L Harris. Antiangiogenic therapy in oncology: current status and future directions. The Lancet, 388(10043):518–529, 2016.

[10] Balaji Ganeshan and Kenneth A Miles. Quantifying tumour heterogeneity with CT. Cancer imaging, 13(1):140, 2013.

[11] Hugo JWL Aerts, Emmanuel Rios Velazquez, Ralph TH Leijenaar, Chintan Parmar, Patrick Grossmann, Sara Carvalho, Johan Bussink, René Monshouwer, Benjamin Haibe-Kains, Derek Rietveld, et al. Decoding tumour phenotype by non-invasive imaging using a quantitative radiomics approach. Nat. Commun., 5(4006):1–8, 2014.

[12] S Baiocco, A Bevilacqua, D Barone, A Delmonte, and G Gavelli. L'eterogeneità perfusionale: un biomarker prognostico per il cancro del polmone non a piccole cellule (NSCLC). Il Giornale Italiano di Radiologia Medica, 4(4):665–669, 2017.

[13] Lejla Alic, Wiro J Niessen, and Jifke F Veenland. Quantification of heterogeneity as a biomarker in tumour imaging: a systematic review. PloS one, 9(10):e110300, 2014.

[14] RCJ Beckers, S Trebeschi, M Maas, RS Schnerr, JML Sijmons, GL Beets, JB Houwers, RGH Beets-Tan, and DMJ Lambregts. CT texture analysis in colorectal liver metastases and the surrounding liver parenchyma and its potential as an imaging biomarker of disease aggressiveness, response and survival. European journal of radiology, 102:15–21, 2018.

[15] Renaud Lopes and Nacim Betrouni. Fractal and multifractal analysis: a review. Medical image analysis, 13(4):634–649, 2009.

[16] Sugama Chicklore, Vicky Goh, Musib Siddique, Arunabha Roy, Paul K Marsden, and Gary JR Cook. Quantifying tumour heterogeneity in 18 F-FDG PET/CT imaging by texture analysis. European journal of nuclear medicine and molecular imaging, 40(1):133–140, 2013.

[17] Zhaohai Yang, Laura H Tang, and David S Klimstra. Effect of tumour heterogeneity on the assessment of ki67 labelling index in well-differentiated neuroendocrine tumours metastatic to the liver: implications for prognostic stratification. The American journal of surgical pathology, 35(6):853–860, 2011.

[18] B Ganeshan, K Skogen, I Pressney, D Coutroubis, and K Miles. Tumour heterogeneity in oesophageal cancer assessed by CT texture analysis: preliminary evidence of an association with tumour metabolism, stage, and survival. Clinical radiology, 67(2):157–164, 2012.

[19] Connie Yip, David Landau, Robert Kozarski, Balaji Ganeshan, Robert Thomas, Andriana Michaelidou, and Vicky Goh. Primary esophageal cancer: heterogeneity as potential prognostic biomarker in patients treated with definitive chemotherapy and radiation therapy. Radiology, 270(1):141–148, 2014.

[20] Philippe Lambin, Emmanuel Rios-Velazquez, Ralph Leijenaar, Sara Carvalho, Ruud GPM van Stiphout, Patrick Granton, Catharina ML Zegers, Robert Gillies, Ronald Boellard, André Dekker, and Aerts Hugo J. Radiomics: extracting more information from medical images using advanced feature analysis. Eur. J. Cancer, 48(4):441–446, 2012.

[21] Charles A Harlow and Sharon A Eisenbeis. The analysis of radiographic images. IEEE Transactions on Computers, 100(7):678–689, 1973.

[22] S Herlidou-Meme, JM Constans, B Carsin, D Olivie, PA Eliat, L Nadal-Desbarats, C Gondry, E Le Rumeur, I Idy-Peretti, and JD De Certaines. MRI texture analysis on texture test objects, normal brain and intracranial tumours. Magnetic resonance imaging, 21(9):989–993, 2003.

[23] Davide Prezzi, Aisha Khan, and Vicky Goh. Perfusion CT imaging of treatment response in oncology. European journal of radiology, 84(12):2380–2385, 2015.

[24] G Petralia, L Bonello, S Viotti, L Preda, G d'Andrea, and M Bellomi. CT perfusion in oncology: how to do it. Cancer Imaging, 10(1):8–19, 2010.

[25] Nunzia Tacelli, Teresa Santangelo, Arnaud Scherpereel, Alain Duhamel, Vale´rie Deken, Ernst Klotz, Alexis Cortot, Jean-Jacques Lafitte, Frédéric Wallyn, Jacques Remy, et al. Perfusion CT allows prediction of therapy response in non-small cell lung cancer treated with conventional and anti-angiogenic chemotherapy. Eur. Radiol., 23(8):2127–2136, 2013.

[26] Gaspard d'Assignies, Anne Couvelard, Stéphane Bahrami, Marie-Pierre Vullierme, Pascal Hammel, Olivia Hentic, Alain Sauvanet, Pierre Bedossa, Philippe Ruszniewski, and Valérie Vilgrain. Pancreatic endocrine tumours: tumour blood flow assessed with perfusion CT reflects angiogenesis and correlates with prognostic factors. Radiology, 250(2):407–416, 2009.

[27] Avinash Kambadakone, Sam S Yoon, Tae-Min Kim, Daniel L Karl, Dan G Duda, Thomas F DeLaney, and Dushyant V Sahani. CT perfusion as an imaging biomarker in monitoring response to neoadjuvant bevacizumab and radiation in soft-tissue sarcomas: comparison with tumor morphology, circulating and tumor biomarkers, and gene expression. Am. J. Roentgenol., 204(1):W11–8, 2015.

[28] Alessandro Bevilacqua, Domenico Barone, Serena Baiocco, and Giampaolo Gavelli. A novel approach for semi-quantitative assessment of reliability of blood flow values in DCE-CT perfusion. Biomed. Signal Process. Control, 31:257–264, 2017.

[29] Serena Baiocco, Domenico Barone, Giampaolo Gavelli, and Alessandro Bevilacqua. Analysis of CT Perfusion Blood Flow maps in Patients with Lung Cancer: Correlation with the Overall Survival. In 23rd Conference of Open Innovations Association (FRUCT), pages 55–61. IEEE, 2018.

[30] Serena Baiocco, Domenico Barone, Giampaolo Gavelli, and Alessandro Bevilacqua. Texture Analysis of Non-Small Cell Lung Cancer on Unenhanced CT and Blood Flow Maps: a Potential Prognostic Tool. In 24th Conference of Open Innovations Association (FRUCT), pages 31-39. IEEE, 2019.

[31] Yoshiharu Ohno, Yasuko Fujisawa, Hisanobu Koyama, Yuji Kishida, Shinichiro Seki, Naoki Sugihara, and Takeshi Yoshikawa. Dynamic contrast-enhanced perfusion area-detector CT assessed with various mathematical models: Its capability for therapeutic outcome prediction for non-small cell lung cancer patients with chemoradiotherapy as compared with that of FDG-PET/CT. Eur. J. Radiol., 86:83–91, 2017.

[32] KA Miles and MR Griffiths. Perfusion CT: a worthwhile enhancement? Br. J. Radiol., 76(904):220–231, 2003.

[33] KA Miles. Perfusion CT for the assessment of tumour vascularity: which protocol? Br. J. Radiol., 76(1):S36–42, 2003.

[34] Gunnar Brix, Stefan Zwick, Jürgen Griebel, Christian Fink, and Fabian Kiessling. Estimation of tissue perfusion by dynamic contrast-enhanced imaging: simulation-based evaluation of the steepest slope method. Eur. Radiol., 20(9):2166–2175, 2010.

[35] Sylvain Goutelle, Michel Maurin, Florent Rougier, Xavier Barbaut, Laurent Bourguignon, Michel Ducher, and Pascal Maire. The Hill equation: a review of its capabilities in pharmacological modelling. Fundam. Clin. Pharmacol., 22(6):633–648, 2008.

[36] Alessandro Bevilacqua, Giampaolo Gavelli, Serena Baiocco, and Domenico Barone. CT Perfusion in Patients with Lung Cancer: Squamous Cell Carcinoma and Adenocarcinoma Show a Different Blood Flow. BioMed Research International, 2018, 2018.

[37] Serena Baiocco, Bert-Ram Sah, Andrew Mallia, Christian Kelly-Morland, Radhouene Neji, James Stirling, Sami Jeljeli, Alessandro Bevilacqua, Gary JR Cook, Vicky Goh. Exploratory radiomic features from integrated 18F-fluorodeoxyglucose positron emission tomography/magnetic resonance imaging are associated with contemporaneous metastases in oesophageal/ gastroesophageal cancer, European journal of nuclear medicine and molecular imaging, 46(7):1478–1484, 2019.

[38] Robert M Haralick, Karthikeyan Shanmugam, et al. Textural features for image classification. IEEE Transactions on systems, man, and cybernetics, (6):610–621, 1973.

[39] E Roy Davies. Machine vision: theory, algorithms, practicalities. Elsevier, 2004.

[40] Sergios Theodoridis and Konstantinos Koutroumbas. Pattern Recognition, Second Edition. Academic Press, Inc., Orlando, FL, USA, 2003.

[41] Elisa Meldolesi, Nicola Dinapoli, Roberto Gatta, Andrea Damiani, Vincenzo Valentini, and Alessandra Farchione. How Can Radiomics Improve Clinical Choices? In Multidisciplinary Management of Rectal Cancer, pages 135–149. Springer, 2018.

[42] Starkov P, Aguilera TA, Golden DI, Shultz DB, Trakul N, Maxim PG, Le QT, Loo BW, Diehn M, Depeursinge A, Rubin DL. The use of texture-based radiomics CT analysis to predict outcomes in early-stage non-small cell lung cancer treated with stereotactic ablative radiotherapy. The British journal of radiology. 2019, 92(1094):20180228.